PROLEGOMENA MATHEMATICA

PHILOSOPHIA ANTIQUA

A SERIES OF STUDIES
ON ANCIENT PHILOSOPHY

FOUNDED BY J.H. WASZINK† AND W.J. VERDENIUS†

EDITED BY

J. MANSFELD, D.T. RUNIA
J.C.M. VAN WINDEN

VOLUME LXXX

JAAP MANSFELD

PROLEGOMENA MATHEMATICA

PROLEGOMENA MATHEMATICA

FROM APOLLONIUS OF PERGA TO LATE NEOPLATONISM

With an Appendix on
Pappus and the History of Platonism

BY

JAAP MANSFELD

BRILL
LEIDEN · BOSTON · KÖLN
1998

This book is printed on acid-free paper.

Library of Congress Cataloging-in-Publication Data

Mansfeld, Jaap.
  Prolegomena mathematica : from Apollonius of Perga to late
Neoplatonism : with an appendix on Pappus and the history of
Platonism / by Jaap Mansfeld.
    p.   cm. (Philosophia antiqua, ISSN 0079-1687 ; v. 80)
  Includes bibliographical references (p.   -) and indexes.
  ISBN 9004112677 (acid-free paper)
  1. Mathematics, Greek.   I. Title.   II. Series.
QA22.M34   1998
510'.938—dc21   98-38382
                                                          CIP

Die Deutsche Bibliothek - CIP-Einheitsaufnahme

Mansfeld, Jaap:
Prolegomena mathematica : from Apollonius of Perga to late
Neoplatonism. With an appendix on Pappus and the history of
Platonism. By Jaap Mansfeld. – Leiden ; Boston ; Köln : Brill, 1998
  (Philosophia antiqua ; Vol. 80)
  ISBN 90-04-11267-7

ISSN 0079-1687
ISBN 90 04 11267 7

PRINTED IN THE NETHERLANDS

TABLE OF CONTENTS

PREFACE

This little book grew out of a paper I was invited to write for a *Festschrift*. Because, alas, things got out of hand I have to publish the results of my enquiries separately. I worked on it from September 1997 to February 1998, adding the indexes later, at the proof stage, and making a few small changes at the same time. One may note that *ANRW* II.37.5, which according to the announcement is to contain a number of survey chapters on ancient mathematics, will appear only a few years from now. This is a pity, because the contents of this volume certainly would have been a great help. The reason why I started working on this theme at all is that I discovered that in an earlier book I had overlooked quite a lot of important evidence, as is explained in the first paragraph on p. 1 below.

A short version of Appendix 2 was delivered as a *Mededeling* (lecture) at the Netherlands Royal Academy on March 9 1998; a longer version, based on a hand-out containing the more important texts, was presented in the context of the séminaire *Les philosophes et la philosophie* at the Sorbonne on March 26 1998. David Runia persuaded me to include a revised English version of this piece in the book. I hope to have profited from the critical remarks made at these oral presentations.

Thanks are due to friends and colleagues who helped in various ways. Keimpe Algra, Pierluigi Donini, Tiziano Dorandi, Frans de Haas, David T. Runia, and Carlos Steel commented on and criticized draft versions, including that of Appendix 2. Petri Mäenpää kindly sent me a copy of his important dissertation on Analysis, a difficult topic on which we also exchanged e-mail letters. Needless to say I take full responsibility for such errors as undoubtedly remain. Henri van de Laar weeded out typing errors and gave indispensable assistance with the bibliography and proofs. My student assistants Ivo Gerardts and Johannes Rustenburg indefatigably brought the books and journals I needed from the University Library. Gonni Runia with her usual expertise again gave the finishing touches to the camera-ready copy.

Bilthoven, July 1998

CHAPTER ONE

PRELIMINARIES

As I discovered to my embarrassment when it was too late, I failed
to include most of the rich evidence available in the fields of
ancient mathematics, both pure and applied, and mathematical
astronomy, in my study of the so-called isagogical questions and
some further, related issues in ancient commentaries, introductions, autobibliographies, and similar literature.[1] (It should be kept
in mind that astrology, not always rigorously distinguished from
astronomy in the modern way,[2] was viewed as a mathematical
subdiscipline.)[3] However this omission—which as far as I know

---

[1] Mansfeld (1994), though I mentioned in passing Theon of Smyrna's
Expositio rerum mathematicarum ad legendum Platonem utilium, and discussed at
some length Proclus' Commentary on Euclid *Elements* I and the traditions
concerned with Aratus (including Hipparchus). On Proclus on Euclid I have
little to add, and on the *Aratea* nothing. No mathematical or mathematico-
astronomical literature is listed in the *apparatus superior* of the first pages of
the edition of Stephanus by Westerink (1985) or mentioned in Hadot (1990a).
Though much has been lost, what has been preserved is impressive, and
without doubt I have missed some things. Diophantus has been excluded
because he has nothing to offer in our present context. Succinct and very
informative (though naturally not up-to-date) overview of ancient authors and
modern editions at Devreesse (1954) 233-43 (mathematics, mechanics,
astronomy), 244-5 (canonics), 252-4 (astrology). Apart from Euclid and Heron
of Alexandria the mathematicians and astronomers are not yet available in
the *TLG.*
[2] Ptolemy for instance in the introduction to the *Apotelesmatica* argues that
these are equally scientific disciplines concerned with foreknowledge in
relation to the heavenly bodies; see below, Ch. IX 2. See further e.g. Lloyd
(1987) 43. Yet it is not my intention to include more than a few samples from
the vast astrological literature.
[3] It is of some interest to quote Simpl. *in Phys.* 293.11-6 Diels: 'the ancients
applied the term 'astrology' to what is now called 'astronomy', because it
would seem that the art of fortune-telling had not yet arrived in Greece. Later
generations made a terminological distinction, applying the name 'astronomy' to the discipline which studies the motions of the heavenly bodies, and
giving the specific name 'astrology' to the art which busies itself with the
effects of these motions on human destiny' (τὸ τῆς ἀστρολογίας ὄνομα οἱ μὲν
παλαιοὶ μήπω τότε τῆς ἀποτελεσματικῆς εἰς τοὺς Ἕλληνας, ὡς ἔοικεν, ἐλθούσης ἐπὶ τῆς
νῦν καλουμένης ἀστρονομίας ἔφερον, οἱ δὲ νεώτεροι διελόντες τοὔνομα τὴν μὲν τὰς
κινήσεις τῶν οὐρανίων ἐπισκοποῦσαν ἀστρονομίαν καλοῦσι, τὴν δὲ περὶ τὰ
ἀποτελούμενα ἐξ αὐτῶν διατρίβουσαν ἀστρολογίαν ἰδίως ἐπονομάζουσι.)

has not been noticed by reviewers[4]—allows me to play Jekyll to my own Hyde, since one of the aims of my earlier study was to try and find antecedents in earlier (even very much earlier) works for the explicit scholastic introductory scheme, the *accessus ad auctores* as it was called in medieval times, of the late Neoplatonist commentators.

As is well known, mathematics and astronomy were taught in the philosophical establishments of late antiquity; names that come to mind are Hypatia, Proclus, Ammonius Hermiae, Marinus of Neapolis, and Simplicius. An investigation of the various kinds of mathematical literature that are involved not only enables one to include the evidence in these fields relating to late antiquity, but also to look for earlier antecedents. As it is, insofar as the isagogical questions are concerned these other traditions (if that is what they may be called) provide a number of excellent parallels to those in the fields of philosophy, belles-lettres, medicine, biblical studies, rhetoric,[5] and grammar. The evidence that is available shows that the study and teaching of mathematics, from the Hellenistic period onwards at least, was not an isolated affair but is to be understood as being a part of the same cultural traditions as the study and teaching of these other disciplines.

With two exceptions[6] the mathematical traditions have not been studied from the vantage point of the present enquiry. I shall attempt to deal with original authors such as the great mathematician Apollonius of Perga (3rd/2nd cent. BCE), and the astronomical works of another great man, the philosophically inclined mathematical polymath Ptolemy of Alxandria (2nd cent. CE), both of whom make use of isagogical questions in an implicit way that is nevertheless unmistakable. Heron of Alexandria (mid-1st cent. CE) was a prolific and technically very competent author in several fields of applied mathematics, and an author of introductory treatises;[7] in these capacities he, too, raises isagogical issues.

---

[4] Chiaradonna (1997) in his review points out important passages in Plotinus and Porphyry which had escaped me, and so corrects another mistake by clarifying the position of the latter.

[5] Rabe's *Prolegomenon Sylloge* with its important introduction has been reprinted in 1995. See forther below, p. 122, complementary note 5.

[6] Schissel von Fleschenberg (1930), though to a certain extent only, see below, nn. 202 and 250; Mogenet (1956) is almost entirely correct, see below, Ch. X 3.

[7] For another work, viz. his Commentary, or comments, on Euclid's *Elements* see below, Ch. III 1.

He also wrote comments, or a Commentary, on the *Elements*. The Neopythagorean Nicomachus of Gerasa (later 1st or earlier 2nd cent. CE) made quite a splash with his Platonizing *Introductio arithmetica*, and he and his commentators, both known (one of them being Iamblichus) and anonymous, are also of some importance. Pappus of Alexandria (first part of 4th cent. CE), more an inventive and competent teacher of mathematics than an original mathematician but also a person with an interest in philosophy, in his wideranging *Collectio* takes these introductory issues into his stride more or less implicitly as well, and does so quite explicitly in his Commentary on Euclid's *Elements* book X. Part of Pappus' Commentary on Ptolemy's *Mathèmatikè Suntaxis*, or *Almagest*, is extant, and in book VI of the *Collectio* he deals with other astronomical works. I shall also look at Theon of Alexandria (mid-4th cent. CE), in his role as editor of Euclid's *Elements* and commentator on Ptolemy. It might be maintained that Pappus and Theon form a sort of Alexandrian succession (*diadochè*), though not necessarily in an institutional sense[8] (there is, at least, no evidence for this assumption). The commentaries on three works of Archimedes and on Apollonius *Conica* books I-IV by Eutocius of Ascalon (early 6th cent. CE), one of the numerous pupils of Ammonius Hermiae (ca. 440-520), will also be included. So will the introduction to Euclid's *Data* of Proclus' pupil and successor Marinus of Neapolis' (5th-6th cent. CE), as well as several anonymous pieces: a substantial introduction to Euclid's *Optica* which has been attributed by scholars to Theon, a late introduction to and commentary on the first book of Ptolemy's *Suntaxis*, and late Prolegomena to Nicomachus' *Introductio*.

As is obvious this approach will also involve, as a side issue, the relation between philosophy and mathematics, but will do so from the point of view of mathematics and its subdisciplines, not from that of philosophy.[9]

---

[8] Both taught the *Elements* and the *Suntaxis*, and Theon used Pappus in his Commentary on the *Suntaxis*. We should perhaps include Serenus of Antinoupolis, of uncertain date but perhaps to be dated before Pappus, even as early as the beginning of the 3rd cent. CE. See Decorps-Foulquier (1992) 56-7, who quotes an anonymous note in *Par. gr.* 1918: in utterances on Plato he allegedly was influenced by the Middle Platonist Harpocration. Serenus wrote a sort of supplement to Apollonius' *Conica* and a (lost) Commentary on Apollonius' grand treatise, see below, n. 142.

[9] For philosophy and mathematics from the point of view of philosophy see Hadot (1984) 379, index s.v. 'mathématiques' (but cf. below, n. 325 and text

For reasons which hopefully will become clear in the course of this enquiry I shall not adhere to a rigid historical or systematic order, but begin with Pappus in whose major work several branches of mathematics both pure and applied, are dealt with. More strictly mathematical literature is next, followed by astronomical literature, though treatment of Ptolemy, as already of Heron, will necessitate that of other mathematical subdisciplines, or disciplines applying mathematics, as well. I finally revert to mathematics, that is to say to the arithmetic, or theory of numbers, of Nicomachus of Gerasa and to those who used it or wrote about it.

A brief reminder: according to the late systematics the main questions to be settled, or at least discussed, before the study of an author, or a text, are roughly as follows.[10] (1) The theme, aim or purpose (σκοπός, πρᾶγμα or ὑπόθεσις) of a particular work, also designated the intention or project (πρόθεσις) of the author or his book;[11] this may include a *historical* excursus, i.e. a discussion of predecessors in the same field or genre, or on the same theme. (2) Its position in a corpus of writings, which involves the further issue of the systematic ordering (τάξις) of such a corpus which may or may not be the same as the most advisable order of study (τάξις τῆς ἀναγνώσεως). Such a τάξις may also apply to the contents of an individual work. (3) Its utility (χρήσιμον, ὠφέλεια etc.)[12] (4) The explanation of its title (αἴτιον τῆς ἐπιγραφῆς). (5) The issue of its

---

thereto), and especially the excellent study of O'Meara (1989), see below, p. 130, complementary note 308.

[10] See also below, p. 122, complementary note 5. Succinct and detailed overview restricted to the late commentators on Aristotle and Porphyry's *Isagogè* at Westerink (1990) 341-8; see further Mansfeld (1994) 192-3 on other secondary literature, 241-3 index *s.v.* isagogical questions, and 195-7 for precedents in earlier authors, esp. Aristotle. As Pierluigi Donini points out to me, in the earlier book I neglected to quote an important passage, Arist. *EN* 1.3.1095a11-3, the summary of the preceding three chapters: 'this much, by way of a *proem*, about the *student*, about *how* (what we say) should be *understood* [this concerns the manner of presentation, cf. below, p. 128, complementary note 217] and our *aim*', καὶ περὶ μὲν ἀκροατοῦ, καὶ πῶς ἀποδεκτέον, καὶ τί προτιθέμεθα, πεφροιμιάσθω τοσαῦτα.

[11] See below, pp. 122-3, complementary note 11.

[12] In our present context, it is worth recalling that Plato in his account of the five mathematical sciences, *Resp.* 7.522c ff., emphasizes their indispensability and utility both for turning the soul towards the intelligibles and in everyday life. Nicomachus often refers to this passage, e.g. *Ar.* 1.1.3 at 8.8-9.4 Hoche. Numerous similar references in Theon of Smyrna, e.g. *Util.* 2.14, 3.7, 5.11, 6.12, 16.4 Hiller. Arist. *Met.* A 1.981b14 ff. distinguishes the productively useful (χρήσιμον) aspect of the arts and sciences from the cognitive. Cf. also below, n. 72.

authenticity (γνήσιον). (6) Its division into parts, e.g. books or chapters (διαίρεσις or τομὴ εἰς κεφάλαια or τμήματα or μέρη). (7) The question to what section of a particular (sub)discipline or literary (sub)genre it belongs (ὑπὸ ποῖον μέρος ... ἀνάγεται). (8) The clarity or lack of clarity (ἀσάφεια) of the author, or of the text, and the reasons for this ἀσάφεια, which is linked to the issue of the manner of presentation, or teaching (τρόπος τῆς διδασκαλίας). (9) The qualities required of the student, and/or of the teacher. (10) In the case of a canonic corpus, e.g. that of Plato: what is the first work to be studied.

It goes without saying that whenever we are dealing with the proem to a work composed by its author the question of authenticity does not arise. It also is true that other preliminary issues, such as e.g. the contents, or division into parts, of a work may be treated in the proem(s) or elsewhere in the work by the author himself. As a matter of fact, the proportion of original authors to commentators or summarizers to be discussed in the following pages is about equal.

Finally, reference to preliminary questions will be effected by *italicizing* the formulas, or notions, that are involved, or by quoting the Greek (technical) terms. For cataloguings of these terms the reader is referred to the Index rerum et nominum antiquorum.

My book of 1994 was criticized by Tarrant (1995). This is not the place for a full reply. I prefer to address briefly his main point, viz. that I failed to acknowledge that introductions to texts, or authors, were also written to further reading, or study, without the help of a master. I have no wish to deny (and never did) that people read things on their own; they certainly did so later in life. Even so, the great literary classics were first read under the direction of the *grammatikos*; later on one could read them for pleasure, and on one's own. Anyhow difficult and technical, or controversial, subjects were, I would maintain, invariably studied under the direction of a master, at least in the earlier stages of one's education. It does not matter whether the instruction was given to a single, so to speak private pupil—as e.g. Crassus is said to have read a not too difficult dialogue of Plato under the direction of a professional, Cic. *De or.* 1.47, 'cum Charmada diligentius legi Gorgiam'—or to a group of students. Medicine, rhetoric, philosophy, and (as we shall see *ad satietatem*) mathematics simply had to be learned with the help of an expert teacher, who of course could write his own textbook (or use one written by someone else) to offer further assistance to people doing their home-work.[12a]

[12a] See now also Barnes (1997) 48 ff.

CHAPTER TWO

PAPPUS' COLLECTIO

II.1 *Introduction*

The Greek text of Pappus' Συναγωγή[13] in eight books lacks book I, a substantial part of book II, and the end of book VIII (extant completely only in an Arabic translation), perhaps also the beginnings of other books. It also underwent modifications in the course of its transmission. It is a miscellany: a number of books are about a wide variety of issues concerned with problems in geometry, though the remains of book II are about calculation. Other books are systematic collections of abstracts of earlier mathematical literature (including disciplines such as mathematical astronomy and mechanics), combined with comments by the author of the collection, in particular in the form of introductory discussions or further or (in Pappus' view) better proofs, called λήμματα because of the additional assumption involved.[14] It is priceless because of the information concerning otherwise lost works it provides.[15]

It has been pointed out by scholars, correctly enough, that the individual books are different as to structure and state of perfection (some having a dedication, others lacking one etc.), and hypothesized that the collection was assembled after Pappus' death from his "foul papers", or drafts.[16] This is an attractive thought, but the mere

---

[13] For other examples of Συναγωγή in book-titles ("collection of material" or "*epitome rei tractatae*") see Mansfeld and Runia (1997) 323-4. Also Procl. *in Tim.* 2.76.23-8 Diehl, who promises to provide a συναγωγὴν τῶν πρὸς τὸν Τίμαιον μαθηματικῶν θεωρημάτων culled from Euclid, Archimedes and others at the end of the course (see further below, Appendix 2, p. 115).

[14] This meaning ('*theorema* auxiliare, *quod* ad demonstrandum hoc de quo agitur theorema *adsumitur*', Hultsch [1876-8] 3.2.66, his italics), not in the new LSJ, should not be ꞁnfused with what we are accustomed to call lemma (quoted portion of text) in a commentary. It has been conjecturally restored at Philod. *Ac. hist.* Y 15, see Dorandi (1991a) 209.

[15] Description and analysis of contents at Heath (1921) 2.357-439, Ver Eecke (1933) 1.xiii-cxiv, Ziegler (1949) 1101-6, Bulmer-Thomas (1974) 294-8, Jones (1986a) 1.5-9, 15-23. "Essays on the lost works" treated by Pappus in book VII at Jones (1986a) 2.510-99; for Apollonius see also below, n. 29. For the textual tradition see below, Ch.III 3 *ad init.*

[16] By Jones (1986a) 1.22-4, following Ziegler (1949) 1094-5. Also see Knorr

fact that four books do possess a dedication shows that they were published, or were intended to be published, more or less as they are.[17] What is more, this seems to hold for the work as a whole as well, since Pappus himself in the carefully written proem to book III, dedicated to a certain Pandrosion, speaks of what he will offer ἐν τῷ τρίτῳ τούτῳ τῆς Συναγωγῆς βιβλίῳ, 'in this third book of the *Collection*'. Book III at the very least was planned as the *third* book *of the Collectio*, so the *Collectio* itself was at the very least planned and to a certain extent executed *by the author*. Eutocius cites book VIII as a separate work with an interesting title of its own: Πάππος ἐν Μηχανικαῖς εἰσαγωγαῖς, 'Pappus in his Introduction to Mechanics', *in Arch. De sphaer. et cyl.* 3.70.6 Heiberg, which indicates that it circulated on its own.[18] This title is interesting because it demonstrates that an individual book of the *Collectio*, or rather its published predecessor cited by Eutocius, was seen as providing an introduction to a subpart of mathemathics; accordingly, the same may hold for the other books.

Eutocius' reference happens to be the only clear mention of (a book of) Pappus' *Sunagôgè* in the whole of the extant literature in Greek.[19] That individual books are dedicated to different persons is also a feature of Apollonius' *Conica* in the polished version presented to the general public: books I-III to Eudemus, books IV-VII to Attalus.[20] Books VII and VIII of the *Collectio* are dedicated to the same person, Hermodorus, book III as we have seen to Pandrosion, while book V is dedicated to Megethion.[21]

(1989) 229.

[17] For various forms of 'publication' in antiquity see Devreesse (1954) 76-81, Mansfeld (1994) 245, index *s.v.*, Dorandi (1997a) 10.

[18] Heiberg (1880) 368, Jones (1986a) 1.22.

[19] For Marinus see below, Ch. VIII. Jones (1986a) 1.28-9 on the basis of Eutoc. *in Apoll. Con.* 2.184.21-86.10 Heiberg (on 2.186.1-10 see already the pertinent remarks of Heiberg [1880] 364-6) argues that Eutocius, who mis-reports Pappus' view on Apollonius vis-a-vis Euclid (see below, ch IV 2 *ad init.*), probably had a version of *Coll.* VII different from the one we have. But note that Eutocius' reference is to 'Pappus and some others' (Πάππος καὶ ἕτεροί τινες, 2.186.2), so is not at first hand. Reports tend to loose reliability as they are handed on from one author to another; yet it remains true that some predecessor(s) of Eutocius had seen a version of *Coll.* VII. Jones' argument has been refuted by Knorr (1989) 228, 240-1 n. 22, who demonstrates that Eutocius did not have access to the *Collectio*.

[20] See below, Ch. IV 1.

[21] Ptolemy dedicates the two main parts of the *Suntaxis* to the same person, Syrus, to whom also his other works that bear a dedication are addressed, see below, text after n. 226. Cicero routinely re-dedicates the later books of a

Perhaps the best suggestion is that what we have here are Pappus' *Kleine Schriften*, some among which are dedicated individually to a variety of persons, as assembled by himself but left in a partially unrevised state at his death. The reference to his Commentary on Euclid *Elements* book I[22] shows that book VIII of the *Collectio* was composed, or had begun to be revised, after this Commentary had been published. Another important work, the Commentary on Ptolemy's *Megalè Suntaxis* (below, Ch. X 1), presumably had also been published already. What is anyhow clear is that what we have here is related to Pappus' role as a teacher of mathematics,[23] as for instance the dedication/introduction to book III makes quite plain. Here he complains about pupils of another teacher who have received an instruction that is insufficient, and promises to provide the appropriate remedies. This has to do with the isagogical topic of *the qualities required of the student,* and of the *teacher.*

Pappus demonstrates his familiarity with the literature of his field, though he may be largely dependent on earlier exegetical sources.[24] However it is not the mathematics which interests me here and which would be beyond my competence anyway, but the literary and scholastic side of the work, that is to say the information it gives us about the order and manner in which literature belonging to various branches of mathematics was taught, or could be taught, and more especially about the preliminary issues which play a part in the presentation of this material on paper (i.e. papyrus, or vellum) and, one may presume, in oral versions in class. I shall restrict the enquiry to books VII and VI in that order, in the main concentrating on the introductory sections, and at the end add something about related items found in the other books. I treat book VII before book VI because its rather clear structure helps to understand the less clear structure of the other

---

treatise to the dedicatee of the first book: there is nothing unusual about dedications of individual books of a single treatise to the same person, but something special about dedications to different persons.

[22] Quoted below n. 78 *ad finem.*

[23] Thus e.g. Ziegler (1949) 1086.

[24] Up-to-date account of Pappus in Knorr (1989) 225-45, Pt. 1 ch. 9: "The ancient commentators and their methods: Pappus and Eutocius", who emphasizes Pappus' dependence on earlier commentators. Cf. further below, nn. 39 and 43.

book, and feel in a position to do so because the order of the individual books in the *Collectio* itself is of little relevance.

II 2 Collectio *Book VII*

According to the dedication/introduction addressed to a pupil, Hermodorus, book VII is about the '*part* (of mathematics) called *Analysis* ... in my summing-up', ὁ καλούμενος ἀναλυόμενος (*scil.* τόπος) ... κατὰ σύλληψιν, 2.634.3-4 Hultsch.[25] 'Analysis', which as a matter of fact means Analysis—cum—synthesis, i.e. pertains to both the reductive way backwards ('we call this kind of approach Analysis, as being a solution in reverse', τὴν τοιαύτην ἔφοδον ἀνάλυσιν καλοῦμεν, οἷον ἀνάπαλιν λύσιν, 2.634.17-8) and the apodictic way down, is defined at some length, so what came to be called the σκόπος of this book is implied. Analysis pertains to both only in the sense that sometimes it refers to the combined method of Analysis-cum-synthesis. But of course sometimes it means just Analysis, which is followed by the corresponding synthesis, as in Pappus' description. We are told that it is a technique intended for those who want to be able to solve problems set to them in geometry, but *useful* (χρησίμη, 2.634.7) for this purpose only.[26] It is *subdivided* into *two parts*, viz. a part which 'attempts to find the truth and is called theoretical', and a 'problematic' part (2.634.24-6, ζητητικὸν τἀληθοῦς, ὃ καλεῖται θεωρητικόν—ποριστικὸν τοῦ προτα-θέντος, ὃ καλεῖται προβληματικόν). Following Mäenpää, one may say that Analysis may yield an absurd, i.e. negative outcome; if

---

[25] See Panza (1997) 383-4 on Pappus' expression κατὰ σύλληψιν, which means something like 'as I summarize it'. For the formula ἀναλυόμενος τόπος see Jones (1986a) 2.377-9; the full version is found at 2.672.4 Hultsch, and in Eutocius, see below n. 207. One should add that τόπος is quite common as a designation for a 'part' or 'subpart' of philosophy, see e.g. Janáček (1992) 253-4 *s.v.* Also see Nicom. *Ar.* 2.6.1 at 82.14-5 Hoche on a subpart of arithmetic: τὸν τόπον τοῦτον (cf. below, text to n. 301). Ptolemy refers to the contents of books III-IV of the *Apotelesmatica* as the γενεθλιαλογικὸς τόπος (*Apotel.* 213 Boll and Boer, lines 5-6 of the *apparatus*). Iambl. *in Nicom.* 56.18 Pistelli speaks of the περὶ ἀναλογίων τόπος (cf. Nicom. *Ar.* 2.21.2 at 119.19-20 Hoche, who should have put τόπον in the text). Serenus 120.7 Heiberg states he wants to treat the τόπος concerned with sections through the summit of the cone. And so on.

[26] For Analysis-and-synthesis see below, p. 123, complementary note 26. I note here that Apollonius in the preface to book I of the *Conica* said that book III contains theorems which are *useful* for synthesis (see below, text to n. 126, and Knorr (1986) 292). For the view of Marinus see below, text to nn. 209 and 219.

the end-point of the way backwards is an impossible problem, or absurd theorem, then synthesis is superfluous. But Analysis may also yield a positive outcome, and in that case a complementary synthesis is usually given. If the outcome is not impossible or absurd, the synthesis provides the solution of the problem or the proof of the theorem.

We also hear that this matter was '*written*' (γέγραπται) by three men, viz. Euclid 'the Elementarist', Apollonius of Perga,[27] and the elder Aristaeus—who happens to be the only mathematician of that name known to us (but this is by the way). In other words, the treatises ascribed to these men which are to be summarized and commented upon in what follows are *genuine*, though this is not stated explicitly.

Further, are given the *ordering* (τάξις)[28] of these for the most part lost[29] works; somewhat to our surprise also one by Eratosthenes is listed, viz. at the end (2.636.18 ff.) The *total number of 'books'*, the *titles* of the works to which they belong being given, is thirty-three,[30] as follows: one work by Euclid (the *Data* in one book), four works by Apollonius (the *De rationis sectione*, the *De spatii sectione*, the *De sectione determinata* and the *De tactionibus*, each in two books), one by Euclid again (the *Porismata* in three books), three further ones by Apollonius (the *De inclinationibus*, *De locis planis*, and *Conica*, in two, two and eight books respectively), one by Aristaeus (the *De locis solidis* in five books), again one by Euclid (the *De locis qui sunt*

---

[27] Eutocius *in Apoll. Con.* 2.180.11 ff. Heiberg quotes a proposition and proof given by Apollonius ἐν τῷ ἀναλυομένῳ τόπῳ. This vague reference presumably pertains to one of the other works by Apollonius mentioned by Pappus (see below) as belonging to this τόπος (Heiberg [1880] 368 suggests the *De locis planis*; see now Jones [1986a] 2.543-4, who argues that the fragment preserved by Eutocius derives from this work), not to a treatise by Apollonius entitled Ἀναλυόμενος τόπος. What may have happened is that Eutocius found the fragment without book-title (but perhaps with the indication ἐν ἄλλοις δέ φησι) in the margin of one of the copies of the *Conica* he consulted (cf. below, n. 39 and text thereto) and correctly inferred that it belonged with the analytical corpus.

[28] Also see the remark on τάξις at 2.672.4-14, on which Knorr (1986) 217-8.

[29] Lost: see above, n. 15; for Apollonius see also Hogendijk (1986). The *De rationis sectione* is extant in Arabic in two 13th. cent. mss. (Jones [1986a] 510-1; also cf. Bellosta [1997]). The Oxford ms. was edited and translated by Halleius (1706); new transl from the two mss. by Macierowski (1989). No critical modern edition exists.

[30] This is because Pappus counts the number of books of the individual treatises. Note that the mss. read 'thirty-two', i.e. 2×4^2; there must be a corruption somewhere, see Jones (1986a) 2.383.

ad superficiem in two books), and finally one by Eratosthenes (the *De medietatibus* in two books). Accordingly Pappus is concerned with the ἐπιγραφή (author and title) of the treatises on his list, and in particular with the *number of the parts,* in this case books, into which each treatise is *divided.* Of each of these works and books, he tells his pupil that he has summarized both the *contents* so that they may be be *studied* (τὰς περιοχὰς ... πρὸς ἐπίσκεψιν) and the number of 'dispositions and diorisms and cases' (transl. Jones) contained in each of them. (Diorisms are conditions of solvability of a problem). But he will not summarize, discuss, and comment upon all of them: the last work to be treated will be Apollonius' *Conica* in eight books.[31] Anyhow the *division into subparts* of the individual books of the treatises is also attended to, quite carefully. Euclid's *Data* for instance according to Pappus contains ninety theorems (2.638.1-2). He has moreover added the solutions of the difficulties that remain to be solved without omitting anything, or so he claims.

Note that the order of the titles in the introduction differs to some extent from the actual sequence of the epitomes and of the lemmas.[32] No lemmas are provided for Euclid's *Data*[33] but it comes first on the list, and its epitome is the first to be given. Clearly this work is the *first treatise* of the ἀναλυόμενος τόπος *to be studied.*[34] What is more, in the first sentence of book VII we are told that Analysis 'taken as a whole (is) a special resource that was prepared *after* the composition of the *Common Elements* (μετὰ τὴν τῶν κοινῶν στοιχείων

[31] Thus three works, viz. Aristaeus' *De locis solidis*, Euclid's *De locis qui sunt ad superficiem*, and Eratosthenes' *De medietatibus* are not summarized or provided with lemmas, apart from two lemmas to the *De locis qui sunt ad superficiem* at the end of book VII, 2.1004.16 ff.

[32] Jones (1986a) 2.382.

[33] The reason for this omission probably is that Pappus wrote a separate Commentary on the *Data*, see Marinus *in Eucl. Data* 256.22-5 Menge, ὡς ὁ Πάππος ἱκανῶς ἀπέδειξεν ἐν τοῖς εἰς τὸ βιβλίον ὑπομνήμασιν. Moreover, according to Marinus, *loc. cit.*, Pappus demonstrated there that the 'manner of teaching' (see below, text to n. 217, and p. 128, complementary note 217) of the *Data* is *analytical* (κατὰ ἀνάλυσιν). Perhaps this now lost Commentary was published before book VI of the *Collectio* was composed, but then it is odd that Pappus does not refer to it; perhaps later, to make up for what was left out (on purpose?) in what became this book of the *Collectio*. It may be noted that Jones (1986a) 1.22 speculates that Marinus may be referring to a version of book VI of the *Collectio* different from ours.

[34] "Not surprisingly, the *Data* turns out to be the very first treatise in Pappus' list of works in the "Domain of Analysis"'; thus Jones (1986a) 1.68, whose lack of surprise is based on the fact that the work "codifies the basic definitions and fundamental theorems required for Analysis of problems" (*ibid.*)

ποίησιν), for those who want to acquire a power in geometry that is capable of solving problems set to them' (2.634.4-6, transl. Jones). Although it is not absolutely certain that Euclid's *Elements* are referred to,[35] this is the most likely assumption. At any rate an *order of study* is involved: first the *Common Elements*, then *Analysis*; and the *required type of student* is indicated as well. The *Data*, be it noted, are "most closely connected with the *Elements*" since they are about the subject of books I-VI, plane geometry.[36] So Pappus' formula *Common Elements* is best explained as referring to *Elem.* I-VI, which form the basis both of the other books of the *Elements* and of the *Data*. Hence 'common'.

An ordered corpus of this nature, containing works relating to a specific (sub)discipline, immediately recalls the corpora of works to be studied in a certain order which we know from late antiquity: works written by Aristotle and by Plato, by Hippocrates (or [Hippocrates]) and by Galen. Furthermore, an order of study conforming to their systematic ordering of certain books of the Old Testament was already prescribed by Origen in the 3rd century. Galen himself in the 2nd century distinguished two different orders of study of (selections of) his own works, and so, a bit earlier, did Albinus for the works of Plato in his *Prologos* to the study of that philosopher. Thrasyllus' tetralogical ordering of Plato's dialogues and letters, to be dated to the early decades of the first century CE, is set out according to an order of study which simultaneously (at least for the most part) is a systematic ordering.[37] The most striking parallel of Pappus' list with Thrasyllus' catalogue is that the *number* of items is given: thirty–three (?) by Pappus, thirty-six by Thrasyllus.[38] There is of course also a difference, since Thrasyllus does not count individual books (of the *Politeia* and *Nomoi*) or individual *Letters*. Nevertheless Thrasyllus provides a canon of Plato's works, and it appears that Pappus likewise describes (and summarizes) the *canon* of classical works belonging to the field of Analysis. This

---

[35] Jones (1986a) 2.380.
[36] Thus e.g. Heath (1921) 1.421-59 (esp. 322, "We should naturally expect much of the subject-matter of the *Elements* to appear again in the *Data* under the different aspect proper to that book, and this indeed proves to be the case.")
[37] For these authors and corpora see Mansfeld (1994) 242, index *s.v.* 'order of study'.
[38] Also compare Porphyry's systematic ordering of Plotinus' essays: 6 *Enneads* = 54 treatises.

impression is enhanced by the fact that, as we have seen, he states that his comments will only go as far as Apollonius' *Conica*; that is to say, for some reason or other he intends to omit to discuss a part of the corpus.

Furthermore, as Knorr has pointed out this corpus, dominated by works of Apollonius, contains no summaries of treatises to be dated later than the 3rd cent. BCE, and he argues that Pappus' sources may have been "annotated copies of the works under review". We may compare those used by Ammonius Hermiae's pupil Eutocius for his Commentaries on Archimedes and Apollonius two centuries later; what is more, Eutocius himself tells us that he wrote his comments *in margine*, so in fact followed one of the standard procedures.[39] Knorr's first point, though perhaps formulated in too absolute a way, strengthens our impression that this canon of mathematical classics was established before Pappus' time, though one can hardly put a date to it. Authors such as Geminus come to mind, but there is no proof. The sheer bulk of the writings (especially those of Apollonius) constituting the canon need not have precluded that an edition of the whole corpus was available next to those of individual treatises. Plato's collected works were even larger, and an edition with critical signs is attested; this is perhaps to be dated to the 1st-2nd cent. CE.[40] Pappus' collection of

---

[39] Knorr (1986) 339-41; cf. Knorr (1989) 225-9, 237-9. The practice of writing comments *in margine* is not only attested for the Late Neoplatonists of Alexandria, but also for the Neoplatonist school of Athens founded by Plutarch, see e.g. Marinus, *VPr.* 27. For Boethius' use of copies of works of Aristotle with annotations (of various provenance, a sort of *Mehrmänner-kommentar*) in margine see Shiel (1990). For Eutocius' practice see above, n. 27, and below, n. 141 and text thereto. To the best of my knowledge students of Neoplatonism fail to refer to Eutocius in this context. References to written treatment by predecessors in Pappus e.g. 2.650.2-3, εἰ μή τινες τῶν πρὸ ἡμῶν ἀπειρόκαλοι δευτέρας γραφὰς ... παρατεθείκασιν (viz. in Euclid's *Porismata*), and 2.680.15-6, συγκεχωρήκασι δὲ ἑαυτοῖς οἱ βραχὺ πρὸ ἡμῶν ἑρμηνεύειν [the only time this verb occurs in the *Collectio*] τὰ τοιαῦτα κτλ. Jones (1986a) 2.404 believes that οἱ βραχὺ πρὸ ἡμῶν refers to "writers on algebra", but the term ἑρμηνεύειν militates against this suggestion; I believe that Pappus refers to earlier comments on Apollonius. Also 3.1028.9-10, where Pappus says he will treat theoretical mechanics better than earlier writers (τοῦ παρὰ τοῖς πρότερον ἀναγεγραμμένου [*scil.* λόγου]). For references to predecessors see also below, n. 43. Probably the otherwise unknown Heraclitus quoted 2.782.5 ff. is one of these predecessors; the suggestion of Jones (1986a) 2.436 that this person may be earlier than Apollonius is improvable. For Pappus on 'Nicomachus the Pythagorean and others' see below, text to n. 68. Also cf. below, text to n. 74.

[40] D. L. 3.65-6 and two similar texts are printed and discussed at Dörrie and Baltes (1990) 92-6, 347-56.

abstracts, which provides an analytical corpus in miniature, is of course ideal for preliminary teaching.

II 3 Collectio *Book VI*

Book VI lacks a dedication, and its introduction is far shorter than that of book VII. But this book too is about a *part*, or *section*, of mathematics, viz. mathematical astronomy, or the ἀστρονομούμενος τόπος as it is called at 2.474.2 Hultsch. Obviously this expression is analogical to ἀναλυόμενος τόπος. But Pappus this time fails to provide an *ordered* list of works to be *studied*, though the existence of such a list is implied. He complains that those who 'teach' (διδασκόντων) the ἀστρονομούμενος τόπος do so incorrectly, adding comments which are superfluous and omitting comments which are indispensable: the isagogical question of the *qualities* to be expected of a *teacher*. Examples are provided: mistakes of this sort have been made in explaining Theodosius' *Sphaerica*, Euclid's *Phaenomena*, and Theodosius' *De diebus et noctibus*. And these teachers commit the same sort of errors with the other books which *follow* on the list (τῶν ἑξῆς, 2.474.13), as Pappus will demonstrate for each particular case. He discusses selected passages from five or six works, viz. Theodosius' *Sphaerica*, Autolycus' *De sphaera quae movetur*, Theodosius' *De diebus et noctibus*, Aristarchus' *De magnitudinibus et distantiis solis et lunae*, Euclid's *Optica* (perhaps), and Euclid's *Phaenomena*.[41] Thus it would appear that Theodosius' *Sphaerica*, first on the explicit list of three and first to be summarized, is the *first* work to be *studied*.[42]

The remark about these other teachers of mathematical astronomy is of further interest because Pappus clearly refers to *written* sources,[43] i.e. an exegetical *tradition* of sorts concerned with collections of astronomical treatises which in his view calls for improvement. Various collections of such a kind are extant in a

---

[41] Editions: Heiberg (1914), Aujac (1979), Fecht (1927), Heath (1913), Heiberg (1895), Menge (1916). For doubts about Euclid's *Optica* being discussed see Neugebauer (1975) 2.768.

[42] It also is the first item in *Vat. gr.* 204, and in several other mss. (see below, text to nn. 44 and 45). Note that Theodosius has to be dated to ca. 100 BCE, see Neugebauer (1975) 2.749-50.

[43] 'Kommentare zu den Σφαίρικα sind mehrfach benutzt" (Ziegler [1949] 1100); e.g. 2.506.21, ἐνθάδε οἴονταί τινες. For other examples (including annoted texts) see n. 39 above.

number of manuscripts. Some of these include not only the treatises discussed by Pappus but also (in some cases) treatises not mentioned by him, one of them even being Euclid's *Data* which in fact belongs with the ἀναλυόμενος τόπος, whereas in other cases works treated by Pappus in book VI are absent from the mss.[44] In one way or other and to some extent or other these mss. go back to earlier such collections; some are *Sammelhandschriften* with a variety of contents, others contain only a few treatises. This variety, and these differences with what is in Pappus suggest either that in Pappus' days alternative collections existed, or that book VI of the *Collectio* is unfinished (remember moreover that the introduction only lists the titles of half the works that are actually discussed and so merely gives us an impression of what is to follow). But one should not be too sceptical: the first section of the oldest of these mss., *Vaticanus graecus* 204 of the 9th-10th cent., contains a corpus of writings very much resembling that discussed by Pappus.[45]

Furthermore, in the second century CE Galen in ch. 2 of the third book of his *Commentary on Hippocrates' Airs Waters Places* (lost in Greek but extant in Arabic and Hebrew, plus a few Latin fragments) alludes to standard treatises belonging to 'the general category of "spherics"' known to some of the astrologers of Rome. These are identified with some probability by Toomer as Autolycus' *De sphaera quae movetur*, Euclid's *Phaenomena*, and Theodosius' *De diebus et noctibus*, all of which are discussed by Pappus, and extant. We may perhaps also include Theodosius' *Sphaerica*. Galen further mentions by name the astronomers Hipparchus, Dioscurides, and Apollinarius (whose works are lost) who as he tells us have not been studied by the astrologers.[46] These remarks seem to presuppose the existence of a standard corpus (or a least a group of standard elementary treatises) which shared at least three

---

[44] Overview at Mogenet (1950) 165.

[45] See Aujac (1979) 29-30 and above, n. 42. Loria (1914) 494-5 believes that the contents of the corpus could differ from one collection to another, and refers to the corpora in Arabic where this is also the case (cf. below, n. 47). But I fail too see much difference with the varied transmission in Greek. For impressive examples of varied transmissions of (parts of) philosophical corpora see Irigoin (1997) 149-190, for the *corpus hippocraticum ibid.* 191-210.

[46] The chapter in the Arabic translation has been edited, translated into English and commented upon by Toomer (1985); his suggestion that Galen perhaps also alludes to Aratus' *Phaenomena* is less plausible, since he has technical works in mind. An edition with translation of the whole Commentary is being prepared by G. Strohmaier for the *Corpus Medicorum Graecorum.*

titles with Pappus' group, and so provide further support for the assumption of a preliminary astronomical course, or preliminary astronomical reading. The authors (Hipparchus, Dioscurides, and Apollinarius) identified by the learned Galen, two of whom are mere names to us, while with one exception the works of Hipparchus have been lost, may have been added by him from his own vast reading.

Contrary to his procedure at the beginning of book VII, Pappus in book VI omits to give us the exact number of 'books' to be treated. This too either suggests that a plurality of corpora existed at the time, or that book VI is still a draft. But note that the bulk of the canon involved is far smaller than that of the canon of Analysis: all these treatises are short, so a count of 'books' is less necessary than in the case of the huge body of treatises constituting the canon of Analysis. However this may be, that one or more collections of astronomical treatises were taught in Alexandria by the fourth cent. CE is put beyond doubt because of Pappus' reference to those who did teach them.

A subsidiary problem is caused by the title of book VI in the mss., and by a scholion to this title (note that these are additions by a later hand in the oldest ms., *Vaticanus graecus* 218 of the 10th cent., and that the other Pappus mss. are its descendants). It states that the book contains solutions of 'what is in the small astronomical [?]', ἐν τῷ μικρῷ ἀστρονομουμένῳ. Scholars have suggested that the substantive τόπῳ should be supplied with the participle ἀστρο-νομουμένῳ, and argued that "Little Astronomy" or "Small Astro-nomical <Locus>" was the title of the corpus that is still extant in various forms in the mss. tradition, and discussed by Pappus in book VI.[47] The designation would have been given to distinguish this corpus from the 'Big Astronomy', i.e. the *Almagest* (*Mathèmatikè*

---

[47] E.g. Heath (1913) 317-8, Knorr (1989) 698. Mogenet (1950) 162-6 remains sceptical as to this designation, but in the end does not exclude the existence of a corpus. Pingree (1968) 15-6 looks at most of the evidence (including that in the Arabic sources) and argues that the later and larger collections may be based on that known to Pappus. Neugebauer (1975) 2.768-9 is strongly opposed to what—in spite of the Arabic evidence—he calls "a story invented by Vossius". Possibly his stance is influenced by his judgement about the "rather modest quality" of *Coll.* VI, which would be "the outcome of a superficial reading of his [viz., Pappus'] sources" (*ibid.*, 767-8). No doubt whatever at Jones (1986a) 2.378. On the existence of corpora of "Dramendichtern, Rednern und Historikern" see Dorandi (1997a) 15-6, with references to the secondary literature.

Suntaxis) of Ptolemy, which purportedly was to follow in the order of study.

The title of the corpus is also quoted in a Commentary of sorts on the first book of the *Mathèmatikè Suntaxis*,[48] part of which was first published by Hultsch at vol. 3.1138-65 of his edition of Pappus as *Anonymi commentarius De figuris planis isoperimetricis*.[49] At 1142.11 we read: δέδεικται μὲν Θέωνι ἐν τῷ ὑπομνήματι τοῦ Μικροῦ ἀστρονόμου, 'has been proved by Theon in his Commentary on the Little Astronomer'. But Mogenet has seen that the sentence quoted by the anonymous author is found in ch. 3 of the first book of Theon's Commentary on the *Suntaxis*, viz. at 358.1-2 Rome.[50] It is therefore entirely doubtful that Theon wrote a Commentary on (the whole of) the *Little Astronomer*, though this possibility is not rejected out of hand by Knorr.[51] But how, believing Mogenet is right, are we to explain the mistake?

The title of the *Mathèmatikè Suntaxis*[52] is given as follows in the *Suda* lemma on Ptolemy (Π 3033, 4.254.7-8 Adler): τὸν Μέγαν ἀστρονόμον ἤτοι σύνταξιν.[53] The first of these alternatives makes for a nice contrast with Μικρὸς ἀστρονόμος. Both these designations are confirmed by Cassiodorus in the *Institutiones* (to be dated to the fifties of the 5th cent. CE, consequently much earlier than the *Suda* and presumably not much later than the anonymous Commentary on *Synt.* I). A vast literature on astronomy exists 'in both languages'; the greatest astronomer among the Greeks—and the only astronomer to be mentioned by Cassiodorus—is Ptolemy, 'who

---

48 For more on this tract, to be dated to late antiquity (proved by Mogenet [1956]), see below, Ch. X 2.

49 Discussed by Knorr (1989) 688-751, who *ibid.* 195-201 provides a new critical edition of a section of this part of the text.

50 Mogenet (1956) 38-9, who however provides no explanation of the error. Hultsch (1876-8) 3.1143 n. 2 already thought of a possible confusion ("nisi forte Theonis commentarium in librum Ptolemaei compositionis, id est in μέγαν ἀστρονόμον, per errorem ad μικρόν rettulit"). Theon's text is "Ὅτι δὲ ἡ ΕΛ πρὸς τὴν ΛΜ μείζονα λόγον ἔχει ἤπερ ἡ ὑπὸ ΕΘΛ πρὸς τὴν ὑπὸ ΜΘΛ, δείξομεν οὕτως, Anonymus' (3.1142.9 ff.) "Ὅτι δὲ ἡ ΓΘ πρὸς ΘΚ μείζονα λόγον ἔχει ἤπερ ἡ ὑπὸ ΓΖΘ πρὸς τὴν ὑπὸ ΚΖΘ, δέδεικται μὲν Θέωνι κτλ.

51 Knorr (1989) 698 speaks of "a commentary by Theon", and suggests that the reference is to one on Theodosius' *Sphaerica*. But he fails to deal with Mogenet's argument; the only way out would be to suppose that Theon used the same phrase (and proof) in the hypothetical Commentary on (part of?) the Little Astronomer, so that they occurred in both works.

52 That this is Ptolemy's own title is put beyond doubt by his self-references, see below, n. 224. For this work see further below, Ch. IX 1.

53 For Eutocius' evidence for the second alternative see below, text to n. 63.

published two works (*codices*) on astronomy, of which he called the one the Minor and the other the Major Astronomer' (*quorum unum minorem, alterum majorem vocavit astronomum*). There is a confusion here, since the 'Minor Astronomer' is not by Ptolemy.[54] What is important, however, is that Cassiodorus confirms the alternative title for Ptolemy's great work found in the *Suda* lemma, and knows the title of the 'work' which comes before it, though he does not tell us why it does so. In what follows he also seems to allude to Ptolemy's *Canones*. Actually, these three titles (if Cassiodorus' *canones* is a title) are the only ones cited by him in this chapter.

Accordingly the mistake of the author of the Commentary (or perhaps of a *scriba*) is that he said 'Small' instead of 'Big'. But a confusion of this sort is more plausible if something entitled Μικρὸς ἀστρονόμος really existed, which entails that copies of a corpus (or corpora) entitled to this denomination actually circulated. On the other hand, supposing Knorr is right (which I believe is unlikely) and there is no mistake, we would have direct evidence from a Greek source of the existence of such a corpus. As to this rare type of title, naming a professional rather than a profession, discipline, or subject, we should compare that of a still extant treatise falsely ascribed to Galen, viz. the Εἰσαγωγὴ ἢ ἰατρός, presumably the work entitled Ἰατρός which a friend of Galen's found at a bookseller's.[55] Also think of Cicero's *Orator* and *De oratore*, and Tacitus' *Dialogus de oratoribus*. These titles are close to those of plays: the professional as protagonist, representing the profession.

Evidence is available that the words 'small' and 'big' were applied to treatises concerned with the same subject, and on occasion involved an *order of study*. Of his *Synopsis* of his large work *On Pulses* Galen says that it should be studied before the 'big treatise' (ὅστις ἀναγινώσκει τὸ βιβλίον τοῦτο πρὸ τῆς μεγάλης

---

[54] See the chapter *De astronomia*, Cassiod. *Inst.* 2.7.2, 155-23 ff. Mynors (Migne *PL* 70, 1218AB); his description of the contents of these 'codices' is rudimentary to a degree. The current terms *minor* (ἐλάττων) and *major* (μείζων) are equivalent to μικρὸς and μέγας, or μεγάλος; for μείζων and ἐλάττων = *maior* and *minor* see below, pp. 124-5, complementary note 67. Neugebauer (1975) 2.769 n. 16 oddly supposes that Cassiodorus refers to Ptolemy's minor astronomical works.

[55] Ed.: 14.674-797 Kühn. For Galen's reference see his *De libris propriis* 19.8-9 Kühn = *Scr. min.* 2.90.4-13 Mueller; the text of the σίλλυβος, transmitted here as Γαληνὸς Ἰατρός, should presumably be emended to Γαληνοῦ Ἰατρός (for examples of such titles see e.g. Oliver [1951]). The title Ἰατρός is safe.

πραγματείας κτλ.)[56] Damascius in his biography of Isidorus tells us that Theosebius had 'written a little booklet dealing with the intricate subjects to be found in the Big Politeia'.[57] This surely was an introduction to the *Republic*; Μικρὰ πολιτεία would have been a suitable title for it. Philoponus in the first sentence of his Commentary on Nicomachus' *Introductio arithmetica* says that this work has this title because it comes before the Μεγάλα ἀριθμητικά.[58] Photius too, *Bibl.* cod. 187, 142b Bekker, tells us that Nicomachus' *Introductio* came before the *Theologoumena* (πρὸ ταύτης). The brief anonymous *Prolegomena* to Nicomachus' *Introductio*[59] reports that Nicomachus wrote another arithmetical treatise to which he gave the title Μεγάλη Ἀριθμητικὴ ἤτοι Θεολογούμενα.[60] The same *Prolegomena* moreover also refers to the Μεγάλος ἀστρονόμος, evidently Ptolemy's *Suntaxis*.[61]

In view of these parallels the hypothesis that the Μικρὸς ἀστρονόμος was studied before the Μέγας ἀστρονόμος is plausible enough, although we do not know when these designations, or titles, were first applied, or this order of study introduced (supposing it was introduced). Late antiquity is the most plausible guess, and I would submit that Cassiodorus provides us with a *t.a.q.* Support for this hypothesis about such a scholastic order of study is provided by the fact that it looks like a development of a claim made by Ptolemy himself;[62] quite possibly even Ptolemy already reflects common practice. Eutocius *in Arch. De dimensione circuli* 3.232.15-7 Heiberg refers to the Commentaries of 'Pappus, Theon and several others on the Μεγάλη σύνταξις of Klaudios Ptolemaios',[63] so he at any rate

---

56 See further below, pp. 123-4, complementary note 56.
57 Dam. *Isid.* Fr. 109.12-5 Zintzen *ap. Suda s.v.* Ἐπίκτετος, E 2424 (2.36.7-8 Adler), συνεγράψατο μικρὸν βιβλίδιον περὶ τῶν ἐν Πολιτείᾳ τῇ μεγάλῃ κεκομψευμένων.
58 Quoted from Haase (1982) 401; see further below, Ch. XI 2.
59 See below, Ch. XI 3.
60 Lost, though parts are extant in the collection of excerpts called Θεολογούμενα τῆς ἀριθμητικῆς (this is the title in the mss.) falsely ascribed to Iamblichus; ed. De Falco (1922). In this compilation the title is Θεολογούμενα at 17.14, while passages from book II are quoted with the title Ἀριθμητική at 42.1 ff. and 56.7 ff. De Falco. Abstract at Phot., *Bibl.* cod. 187, who gives the ἐπιγραφή as Νικομάχου Γηρασηνοῦ ἀριθμητικῶν θεολογουμένων βιβλία β', 142b Bekker. This title is probably to be translated 'Arithmetical Theology'; for θεολογούμενα in the sense 'theological doctrines' cf. D. S. 1.23.7, 1.29.6, 1.86.3, 3.61.6, Plu. *de Is.* 367C (*SVF* 2.1093), S.E. *M.* 9.56, to quote only parallels from pagan authors.
61 76.10-4 Tannery.
62 See below, n. 234 and text thereto.
63 On these Commentaries see below, Ch. X 1-2.

already knew the work by the latter title.[64] His contemporary
Asclepius, also a pupil of Ammonius Hermiae, likewise refers to
Ptolemy's work by this title: καὶ ὅσα εἴρηται ἐν τῷ πρώτῳ βιβλίῳ τῆς
Μεγάλης συντάξεως, in Met. 359.32 Hayduck.[65]

The expression ἐν τῷ μικρῷ ἀστρονομουμένῳ in the Pappus mss. is
best explained as a conflation of Pappus' formula ἀστρονομούμενος
τόπος and the designation Μικρὸς ἀστρονόμος.[66] The title Μικρὸς
ἀστρονόμος shows that the corpus discussed by Pappus and also
taught by others, composed of works by various hands, could so to
speak be viewed as a single treatise. In fact adjectives such as μέγας
and μικρός were also occasionally used to distinguish from each
other individual works which otherwise would have had exactly
the same title.[67]

II 4 *Further Evidence from the* Collectio

I conclude this chapter with some further evidence in the *Collectio*
relating to isagogical issues. When a summary of the contents of a
particular work is given what came to be called its σκοπός or
ὑπόθεσις as well as its parts are of course involved, though the terms
themselves are not used (above we have seen that Pappus' term for
summarized contents, occurring a few times, is περιοχή). *Utility* is
mentioned quite regularly, e.g. 1.30.21, ὠφέλεια of book III of the
Collectio. A most interesting remark is found at 3.18, 1.84.1ff.,
'Nicomachus the Pythagorean and some others treated not only
the first three proportions, which are most *useful* (χρησιμοί) for the
study of the ancients (πρὸς τὰς τῶν παλαιῶν ἀναγνώσεις), but also
three others which one finds with the ancients, and younger

---

[64] On the title Μεγίστη σύνταξις (whence *Almagest*, via the Arabic) see
Neugebauer (1975) 2. 836-7.

[65] I note in passing the expression ἐπὶ μεγάλαις συντάξεσιν at Herodianus
1.6.8, meaning 'by large subsidies'.

[66] Thus Jones (1986a) 2.378. Note that ἀστρονομούμενος may mean 'a
practitioner of astronomy' (D.L. 1.34, οἶδε δ' αὐτὸν [*scil.,* Θάλητα] ἀστρο-
νομούμενον καὶ Τίμων—who calls him a σοφὸν ἀστρονόμημα) so comes quite
close to ἀστρονόμος (the middle voice is equivalent to ἀστρονομοῦν, cf. Plato *Tht.*
174a). The *Suda* lemma on Manetho (M 143, 3.318.9 Adler) ascribes to this
author Ἀποτελεσματικὰ δι' ἐπῶν, καὶ ἄλλα τινὰ ἀστρονομούμενα, and *s.v.* Porphyry, Π 2098 (4.178.29-31 Adler) tells us that Porphyry wrote numerous other
works, καὶ μάλιστα ἀστρονομούμενα· ἐν οἷς καὶ Εἰσαγωγὴν ἀστρονομουμένων ἐν
βιβλίοις τρισί (Porph. 418T Smith).

[67] See below, pp. 124-5, complementary note 67.

authors have discovered four more'.[68] This is a sort of collage of two passages of Nicomachus ; among Pappus' 'ancients' is 'the most divine Plato'—1.86.21, in a paragraph crammed with reminiscences of the *Timaeus*—also mentioned by Nicomachus.[69] But the ancients he seems particularly to have in mind (some sleight-of-hand being unavoidably involved) are the mathematicians mentioned in a long *historical excursus* earlier in the same book, 3.7 at 54.20-56.17, which deals with the three kinds of geometric problems distinguished by them that are relevant to the study of proportions. Here we find the names of Eratosthenes, Philon, Heron (cf. also 62.14), Apollonius, Aristaeus, and Nicomedes (cf. also 58.23). Pappus states that to the presentation of the solutions of these men he will add what he has further worked out and perfected himself (56.9-10, μετά τινος ἐμῆς ἐπεξεργασίας).

Compare further 1.304.10, τὸ χρήσιμον καὶ βιωφελές which also holds for mathematics as practised and used by humans;[70] 2.676.1 ff., χρεῖα of book II of Apollonius' *Conica*; 3.1022.3-4, mechanics, the subject of book VIII, is in many ways τῶν ἐν τῷ βίῳ χρήσιμος (cf. 3.1024.12 ff., list of useful mechanical arts),[71] and of major *importance* for physics; 3.1046.26 ff., χρεῖα for mechanics of certain propositions. *Explanation* of the *title* of Apollonius' *Inclinationes* as deriving from one of the things stated in this work, 2.670.7-8, ἐπέγραψαν δὲ ταῦτα Νεύσεις ἀπὸ ἑνὸς τῶν εἰρημένων. *Title* given by Eratosthenes, 2.662.15-6 οἱ δὲ ὑπὸ Ἐρατοσθένους ἐπιγραφέντες Τόποι πρὸς μεσότητας (which moreover entails that the work is *authentic*). *To what (part of a) discipline* another discipline *belongs*: 3.1022.13-24.2: according to the followers of Heron mechanics is divided into two, viz. a theoretical (λογικόν) and a technical (in the sense of

---

[68] For other references to secondary literature in Pappus see above, n. 39. For Pappus on Nicomachus see further below, Appendix 2, pp. 117-9. Note that Ammonius Hermiae called Nicomachus a Platonist, not a Pythagorean, see below, text to n. 314.

[69] For more details see below, Appendix 2, pp. 117-9.

[70] Cf. above, n. 12, below, n. 71 and text thereto.

[71] Cf. text to previous n., and Zeno of Citium's well-known and often echoed definition of *technè* at e.g Olymp. *in Grg.* ch, 12.1, 70.7 ff. Westerink (*SVF* 1.73, where also other parallels; add e.g. Sopater *Schol. ad Hermog. Stat.* 5.4.6-7 Walz, Olymp. *in Grg.* ch. 2.2, David *Prol.* 44.5-6 Busse): Ζήνων δέ φησιν ὅτι τέχνη ἐστὶ σύστημα ἐκ καταλήψεων συγγεγυμνασμένων πρός τι τέλος εὔχρηστον τῶν ἐν τῷ βίῳ. For the very common formula χρήσιμον/μα πρὸς τὸν βίον see e.g. *Dissoi logoi* 90 9.1 DK, 2 p. 416.13-4, Xen. *Mem.* 2.7.7 and 4.3.7, Arist. *EN* 10.1.1172b4-5, *Pol.* 8.2.1337a41, D. S. 1.8.5, Gal. *PHP* 9.2.30, Marc. Aur. 4.29.1.

'applied', χειρουργικόν) part (cf. 1028.4-5),[72] the former consisting of geometry, arithmetic, astronomy and physical theory, the latter of metal-working, house-building, carpentry, and painting.[73] *Clarity*: 3.1028.6-10 Hultsch, Pappus will describe the theorems of mechanics found by the ancients and those added by himself in a more concise and clearer way (συντομώτερον καὶ σαφέστερον ἀναγράψαι) than his predecessors.[74]

---

[72] Cf. Ptolemy's distinction between two ways of practising canonics, *Harm.* 5.25-6 Düring, μόνῃ τῇ χειρουργικῇ χρήσει versus θεωρητικώτερον. Pappus' report of Heron's distinction is mentioned by Fuhrmann (1960) 171-2.

[73] Note that Pappus *disertis verbis* restricts his account to the theoretical part (3.1028.4-10).

[74] For the predecessors see above, n. 39. The formula συντομώτερον καὶ σαφέστερον is already found at Isocr. *Archid.* 24.3, then a few times in late authors. The terms are opposed to each other at Them. *in APo* 1.16-2.4 Wallies.

CHAPTER THREE

COMMENTARIES ON EUCLID, THE SCHOLIA ON EUCLID'S
ELEMENTS AND PAPPUS' COMMENTARY ON BOOK X

III 1 *Comments and Commentaries on the* Elements *and* Data

First, a few remarks on the ancient literature dealing with the
Elements (and *Data*) in order to put Pappus' Commentary on Book X
of the *Elements* and the Scholia in their proper context.

In his Commentary on the first book of Euclid's *Elements*[75]
Proclus several times refers (often in critical terms) to earlier
authors or commentators (παλαιοί, or ἐξηγηταί) dealing with Euclid
or with issues that are relevant to the interpretation of the *Elements*.[76]
Occasionally names are mentioned. Heron[77] and Pappus are cited
several times.[78] The mathematician Geminus (1st cent. BCE/CE)[79]

---

[75] Ed. Friedlein (1873), transl. Morrow (1970).

[76] ἐξηγηταί 189.11-12, 200.11-7, 209.11-3, 328.15-6, παλαιοί 121.12, 144.3,
200.12, 272.19, 396.11-2, 422.25 Friedlein; in general see Heath (1926) 1.33-5.
It should be noted here in passing that already several Epicureans, most
importantly Polyaenus (a distinguished mathematician who came to believe
all mathematics is false, Cic. *Luc.* 106), Demetrius of Laconia, and Zeno of
Sidon (criticized in his turn in a book by Posidonius) dealt critically with
Euclid; see Sedley (1976) 23-4, Angeli and Dorandi (1987), a..d Angeli and
Colaizzo (1979) 64-8, esp. on Zeno Sid. Fr. 27 (= Posid. Fr. 46 + 47 Edelstein-
Kidd; also see useful discussion of the Posidonian texts at Kidd [1988] 1.207-
14), to be found at Procl. *in Eucl.*, who in fact mentions Zeno's name seven
times: 199.15, 200.5-6, 214.18, 215.10, 216.10, 217.10, 218.1, and 'the Epicureans'
in general at 322.5 and 323.4 Friedlein.

[77] See below, pp. 125-6, complementary note 77.

[78] There are four explicit references to Pappus' Commentary to book I:
approving at 189.12 ff., 197.6 ff., 249.20 ff., critical at 429.13 ff. Friedlein: οἱ περὶ
"Ηρωνα καὶ Πάππον should not have appealed to proofs in book VI (but see
Heath [1926] 1.366-8). On interpolations from Pappus' Commentary in the
text of Euclid see Heiberg (1903) 57-8. Pappus himself *Coll.* VIII, 3.1106.13-5
refers to his σχόλιον (i.e. Commentary) on book I of the *Elements*.

[79] From Geminus' reference to Chrysippus cited Procl. *in Eucl.* 385.13 ff.
Friedlein (*SVF* 2.365) or parallels with passages in Cleomedes it does not
follow that he was a Stoic. i.e. the same person as Posidonius' excerptor, or
follower; see further Neugebauer (1975) 2.578-9, also *ibid.* 579-81 for Geminus'
dates. Aujac (1975) xi-xiii attributes the extant *Elementa astronomiae,* the lost *De
Posidonii meteorologica* (striking astronomical fragment via Alexander at
Simpl. *in Phys.* 291.22 ff. Diels) and the lost mathematical work all to the
same person, and edits the passage in Proclus on the division of mathematics

is cited quite frequently,[80] but these references do not derive from a Commentary but from the treatise *On the Ordering of the Mathematical Disciplines* (Περὶ τῆς τῶν μαθημάτων τάξεως), in which also specific mathematical treatises were discussed: Euclid's *Elements*, perhaps the *Data*, and certainly works by Archimedes, and Apollonius on conics.[81] The Neoplatonist philosopher Porphyry (3rd cent. CE), another of Proclus' sources, appears to have written not a Commentary on the whole work but comments on book I, which may have been part of his *Miscellaneous Investigations*.[82] Fragments of these Commentaries and comments are extant also elsewhere,[83] and we even have the whole of Pappus' Commentary, in two books, on book X of the *Elements* in an Arabic translation. This is accessible in an English version which replaces earlier short abstracts in French translation and a complete German translation both based on an unreliable edition of the Arabic text.[84] This, it should be

(but not the other references, or those in Pappus and Eutocius) *ibid.* 114-7. Aujac's view is shared by Dicks (1972) and Crombie (1994) 1.137-8, who translates the fragment found in Simplicius. A *non liquet* seems to be the best option.

[80] The division of mathematics into eight parts by 'Geminus and his followers' is cited 38.4-42.8 Friedlein; two pure disciplines, viz. arithmetic and geometry, and six applied ones, viz. mechanics, astronomy, optics, geodesy, canonics and calculation. Overview of passages in the *in Euclidem* either certainly or perhaps deriving from Geminus at Van Pesch (1900) 112-3, but see e.g. Mueller (1992) xxviii.

[81] Ammon. *in APr.* 5.27-8 Wallies τὴν τοιαύτην ἀνάλυσιν ὁ Γεμῖνος ὁριζόμενός φησιν "ἀνάλυσίς ἐστιν ἀποδείξεως εὕρεσις" suggests that Geminus may have had the *Data* in mind. I quote the title after Pappus, 3.1026.5-9, which includes the reference to Archimedes; other references to Geminus on Archimedes at Eutoc. *in Archim. De plan. aeq.* 3.266.1-2, and to Archimedes, Apollonius and other early mathematicians in an abstract from book VI at *in Apoll. Con.* 2.168.17-170.24 Heiberg, where the title is slightly different (ἐν τῷ ἕκτῳ τῆς τῶν μαθημάτων θεωρίας). Presumably the title as quoted by the meticulous Pappus is the correct one, Eutocius' reference being couched in more general terms (cf. his vague reference to Apollonius, above n. 27). Note that Tittel (1912) 1040-1 argues in favour of Eutocius' title, but his parallel, Cleomedes' title Κυκλικὴ θεωρία, is now rejected in favour of Μετέωρα, see Todd (1990) xx-xxi.

[82] Six explicit and laudatory references to Porphyry: 156.24-27.1 Friedlein = Fr. 257T Smith (ὅσα καὶ ὁ φιλόσοφος Πορφύριος ἐν τοῖς Συμμίκτοις [i.e. the Σύμμικτα ζητήματα] γέγραφεν καὶ οἱ πλεῖστοι τῶν Πλατωνικῶν διατάττονται), 255.12-4 = Fr. 482F Smith, 297.1 ff. = Fr. 483F Smith, 315.11 ff. = Fr. 484F Smith, 323.7 in Fr. 485F Smith (see above), 352.13-4 in Fr. 486F Smith. Proclus sees himself as belonging to a philosophical rather than a mathematical exegetical tradition, cf. O'Meara (1989) 170-1; for Porphyry's influence also Mueller (1987) 311-3.

[83] See e.g. Heath (1926) 1.19-27; Jones (1986a) 2.10-11 on Pappus.

[84] Extracts Woepcke (1856), see quotations at Heiberg (1891-3) 2.120-4; transl. Suter (1922) based on Woepcke (1855), replaced by Thomson (1930).

noted, is not a *commentarius perpetuus*. The first book is a lengthy, mostly philosophical introduction to book X which to some degree is comparable to Proclus' two prolegomena to his Commentary on book I, while selected mathematical observations of a sober nature occupy most of the second book. This treatise is sometimes spurned by historians of mathematics,[85] and practically ignored by historians of philosophy.[86]

We do not know whether Theon of Alexandria wrote a separate Commentary, or comments, on the *Elements*, but do know that he published a revised edition.[87] This revision was to serve the purpose of teaching Euclid in a better way. It is mentioned *Schol. Eucl.* I.2, which reports that 'in certain copies the words "according to Theon's edition" are included in the *title*' (ἔν τισιν ἀντιγραφοῖς πρόσκειται ἐν τῇ ἐπιγραφῇ τὸ ἐκ τῆς Θέωνος ἐκδόσεως); similarly *Schol.* IV.4. In fact a number of mss. that are still extant tell us that they are ἐκ τῆς Θέωνος ἐκδόσεως. What is more, Theon himself refers to it in his Commentary on Ptolemy's *Suntaxis* in such terms that it is clear that his comments (whether original or not) were *part of* the edition: 'this has been *proved by us* in the edition of the *Elements* near the end of book VI' (δέδεικται ἡμῖν ἐν τῇ ἐκδόσει τῶν Στοιχείων πρὸς τῷ τέλει τοῦ ἔκτου βιβλίου, 1.10, 492.7-8 Rome). Possibly such additional proofs were originally written *in margine* and incorporated into the body of the text in a later phase of the transmission. This would explain why Theon's name is absent from Proclus' Commentary.

The work is mentioned under 'Pappus the Greek' at *Fihrist* 7.2, Dodge (1970) 2.642: 'a Commentary on the tenth section of Euclid, in two sections' (cf. Suter [1892] 22). See also Sezgin (1974) 174-6. For the reference to it in the scholia to Euclid's *Data* see below, n. 120. That Pappus also commented on book I of the *Elements* is clear from his own reference (above, n. 78 *ad finem*), and from the quotations in Proclus (above, n. 78 *ad init.*) and Anaritius/an-Nayrizi (below, n. 90 and text thereto); that he commented on book XII appears from a reference in Eutocius, see below n. 103 and text thereto.

[85] E.g. Fowler (1987) 302: "Unfortunately, Pappus' commentary is of little help in understanding *Elements* X." Suter (1922) 11 spoke of "philosophisches Beiwerk", but praised the mathematical sections for their clarity.

[86] But see Burkert (1972) 533, index *s.v.* Pappus; his contention that all the scholia on *Elem.* X derive from Pappus may however be contested.

[87] Heiberg (1925) 15-6; Heath (1926) 1.46-61 and Ziegler (1934) 2077-8, mainly based on Heiberg (1882), the "prolegomena critica" in Heiberg (1888a, repr. Stamatis [1977]), and Heiberg (1903) 52-3; in this later paper Heiberg shows that Theon not only made additions and introduced changes but also followed earlier mss. that already contained interpolations. Also see Dorandi (1994) 306-7, 309. On Theon see further Toomer (1976b) 322.

Proclus was not the last Neoplatonist commentator on Euclid. The introduction to the *Data* by his pupil Marinus of Neapolis (in Palestine) is extant.[88] A number of fragments of Simplicius' Commentary on book I of the *Elements*[89] have survived in the medieval Latin translation of an Arabic Commentary on books I-X ascribed to Anaritius (an-Nayrizi, 10th cent. CE), the first book of which is lost in Arabic.[90] To the best of my knowledge these fragments have received little attention, and I cannot deal with them here. To a large degree this Arabic Commentary is a compilation from Greek sources, otherwise lost, the most important among which are Heron (in books I-IX) and, as already stated, Simplicius (in book I). Proclus is not mentioned. From the Arabic text, which mentions his name here, it is clear that the *quidam* mentioned at Anaritius 37.17 and 38.7 Curtze represent Pappus.[91]

III 2 *The* Scholia in Euclidem: *Proclus, Pappus and Others*

We also have several corpora of scholia to the *Elements*, edited by Heiberg in 1888 and discussed by him in an important monograph published in the same year.[92] Heiberg established that with a few exceptions the scholia on book I belonging to the earliest corpus (called by him *Scholia vaticana*) derive from Proclus' Commentary, abstracted and reworked by an intelligent person, and hypothesized that the scholia to books II-XIII belonging to this corpus had in a similar intelligent way been derived from Pappus' Commentary, since in his view there is no evidence that Proclus wrote on the other books. Heiberg knew Woepcke's French translation of extracts of Pappus' Commentary on book X; using the method of the double column which provides an intuitively

---

[88]   Ed. Menge (1896b); see below, Ch. VIII.

[89]   See below, p. 126, complementary note 89.

[90]   Mentioned at *Fihrist* 7.2, Dodge (1970) 2.635 (cf. Suter [1892] 16). Critical ed. of the whole Latin transl. Curtze (1899), new ed. of books I-IV Tummers (1994), who published a preliminary ed. of book I at (1984) 2.121-90. Arabic text ed. Besthorn and Heiberg (1893-1932). On an-Nayrizi, who belongs with "den bedeutenderen arabischen Mathematikern" see Sezgin (1974) 283-5; on Anaritius see Tummers (1984) 2.103-6.

[91]   Cf. above, n. 83 and text thereto.

[92]   Heiberg (1888b); for additional scholia see below, n. 96 and text thereto. Abstract at Heath (1926) 1.64-74; this account is somewhat out of date because Junge and Thomson (1930) was not yet available.

convincing synoptic overview[93] he proved that several passages in the *Scholia vaticana* on this book indeed correspond to passages in Pappus' work.[94] Junge and Thomson at the end of the introduction to the English translation were able to extend this list to some extent. Comparison of the scholia with the complete text shows that the scholiast not only took liberties with it (perhaps enhanced by later users), i.e. by expanding or shortening it, but also, as in the case of the first and quite substantial scholion to book X, wrote a little essay based on Pappus including virtually verbatim abstracts but in a different sequence than in his source.[95]

Subsequent to his first monograph on and publication of the scholia, Heiberg found further scholia in other mss., one of which is quite early.[96] Here *Schol. vat.* X.62 is ascribed to 'the divine Proclus'.[97] Heiberg concludes that there are two possibilities, of which he prefers the first: (1) the ascription is a guess of a Byzantine scholar; (2) Proclus wrote a Commentary on the whole *Elements*, and the passages in the scholia corresponding to Pappus have reached us via Proclus' Commentary. Eva Sachs preferred the second alternative, but her argument for deriving the *Scholia vaticana* as a whole from Proclus is not good enough.[98] She attributes *Schol. vat.* X.1 and X.135 to Proclus (who she thinks would have used the in her view unreliable Iamblichus) because of a "Zug von pythagoreischem Mystizismus" which as she believes does not fit the sober Pappus.[99] But the passages to which she objects are exactly paralleled in Pappus' Commentary on book X (1 §§ 1 and 9), the full text of which was not known to her. She also finds *Schol. vat.* VII.3 Proclean. This is about the monad in the domains of the gods, of physical objects, and of mathematicals; 'when speaking of a monad in relation to the gods we mean the beginning of each

---

[93] See Mansfeld and Runia (1997) 89-94, 116-20.
[94] Heiberg (1882) 170-1, (1888b) 11-2.
[95] Thomson (1930) 57-8.
[96] Published and discussed Heiberg (1903) 328-33, 334-52.
[97] Text Heiberg (1903) 341, no. 17; discussion *ibid.* 345-6.
[98] Sachs (1917) 71-5, "Proklos und die Euklidscholien"; also cf. *ibid.* 38-9. Her contribution was overlooked by Suter (1922), who p. 78 suggests that certain passages in Pappus' Commentary may derive from Proclus and have been interpolated by the Arabic translator (refuted by Thomson [1930] 40-1); it was also overlooked by Junge and Thomson (1930). More on one of Sachs' points below, text to n. 119 and p. 127, complementary note 119. Also see below, n. 114.
[99] Sachs has overlooked Pappus' reference to Nicomachus, for which see above, text to n. 68, and for more details below, Appendix 2, pp. 117-9.

series' (σειράς, 5.362.12-3 Heiberg). It has to be admitted that the term σειρά for ordered series is without any doubt Neoplatonic. Even so, the germ of the idea behind the scholion can be paralleled from Pappus' Commentary, viz. 1 § 8: 'everything finite is in fact finite only by reason of the finitude which is the first of the finitudes'.[100] It is therefore plausible enough that the scholiast modernized and amplified an idea found in Pappus' Commentary to book VII. If we assume (as Heiberg appears to have done) that a single person is responsible for (the majority of) the *Scholia vaticana*, this person must of course be later than Proclus, excerpted by him for book I. So in all probability he was a minor Late Neo-platonist himself; note that the excerpts that are probably derived from Pappus show symptoms of updating. Marinus, of whom it is said that he excerpted earlier Commentaries on a considerable scale,[101] is a possible but of course entirely hypothetical candidate.

Furthermore, by no means all the *Scholia vaticana* on *Elements* book X correspond to passages in Pappus' extant Commentary. So one can be certain that not all the *Scholia vaticana* to *Elements* II-IX and XI-XII derive from Pappus either. But in view of their contents (quite similar to those scholia which may safely be said to stem in one way or other from his Commentary) and because they are part of the same corpus some of them may well have been excerp-ted from Pappus. Heiberg's hypothesis is simple, and therefore plausible,[102] though it needs to be revised in the manner attempted just now. In our present context it does not matter, moreover, whether or not these have to some extent been brought up to date by someone who found it worth his while to excerpt Proclus for book I (we have just seen one clear instance of such an upgrading). We may add that material deriving from Pappus is also found outside the *Scholia vaticana*: *Schol.* XII.2 is proved to derive from his Com-mentary on this book by a remark in Eutocius.[103]

Do we find mention of isagogical issues in the *Scholia vaticana* to books II-XIII which thus may be ultimately attributed to Pappus?[104]

---

[100]   On this passage see Thomson (1930) 40-1.
[101]   Cf. below, n. 200.
[102]   Cf. Ziegler (1946) 1092, who however does not exclude "Quellen-gemeinschaft".
[103]   Eutoc. *in Arch. De sphaer. et cyl.* 3.28.16-7, εἴρηται δὲ καὶ Πάππῳ εἰς τὸ ὑπόμνημα τῶν Στοιχείων.
[104]   As far as I know no general study of the scholia to Euclid has been made after Heiberg (1888b) and (1903); however the contribution of Junge

Yes we do; (abstracts from) prolegomena to books II-V have been preserved, while as we have seen above the introductory scholion to book X (a little essay) derives from a section of Pappus' Commentary which happens to be extant.

The short *Schol. vat.* II.1 explains both the *utility* (χρήσιμον) and the *subject* or *purpose* (σκοπός) of the book, in this (unusual) order. It is useful for many things, because it is a contribution to stereometry and the theory of planes, helps to solve many problems, 'and contributes not a little to astronomy'.[105] Its subject is the description of straight lines and their parts, which will clarify the irrational divisions of straight lines. Implicitly the place of the contents of this book as a part of the discipline involved, viz. mathematics, is also indicated: the isagogical question to *what section of a particular discipline* or literary genre it belongs. The even shorter *Schol. vat.* III.1 only describes the σκοπός. *Schol. vat.* IV.1 is a bit longer; though lacking the technical isagogical vocabulary, it in fact is about the *order* of theorems and provides a brief overview of the limited contents of the book (i.e. tells us about its σκοπός), and equally implicitly deals to some extent with its *utility*: what is at its end forms 'a contribution to astronomical theory'.[106] The very first word of the quite extensive *Schol.* V.1 is σκοπός: the *subject* of the book is the treatment of mathematical proportions (ἀναλογίαι, a term which subsequently is explained at some length). *Utility* is also mentioned, though again implicitly; we are told (5.280.2-7) that the present book is 'common' (κοινόν) to geometry and arithmetic and 'music' (i.e. canonics) and indeed to mathematics in general, for its proofs do not only fit geometric theorems, but all the disciplines which belong to the science of mathematics. Accordingly, the place of the contents of this book in relation to the discipline involved, viz. mathematics, is also indicated: the isagogical question to *what section of a particular discipline* or literary genre it belongs. 'This is its σκοπός', the scholiast continues, 'but some say that the book is the discovery of Eudoxus,[107] the teacher [*sic*] of Plato' (280.7-9). What is implied by this remark is that nevertheless, in its present and quasi perfect shape, it is correctly attributed

---

and Thomson (1930) 57-8 is indispensable, and useful remarks are scattered in the work of Burkert (1972), see 534, index *s.v.* Scholia in Euclidem (which however fails to list all the passages dicussed).

[105] For Pappus' interest in mathematical astronomy see above, Ch. II 3.
[106] Cf. again above Ch. II 3.
[107] Cf. Burkert (1972) 451 with n. 19.

to Euclid: the isagogical issue of *authenticity*, i.e. the correctness of the ἐπιγραφή.[108] A fourth such issue is discussed, explicitly this time, at the end, viz. the *division into parts* (282.2-10, τῆς τοῦ βιβλίου διαιρέσεως). The book is divided into two parts (διχῇ διῄρηται), the first of which provides the διδασκαλία of the simpler subjects (i.e. the multiples), while the second is more general in character. 'For with each topic, as has been said [viz., in a section of the Commentary that is lost], the presentation of the simple subjects should come first'.[109] This comment recalls Porphyry's justification of his systematic arrangement of Plotinus' treatises at *VP* 24,[110] and so is in fact not only about the *systematic sequence* but also about the *order of study*, while the *manner of presentation* is involved as well. The excerpt ends with the remark that the division of the definitions is like that of the book as a whole, the first group being about parts and multiples, the next dealing with all proportions in general.

Schol. vat. X.1 need not be discussed, as we have Pappus himself.[111] *Schol. vat.* XI.1 lacks technical isagogical vocabulary, but it is about the contents (i.e. σκοπός) of the book, and contains an interesting historical observation, viz. that 'the ancients' distinguished the knowledge of planes from that of solids, 'as Plato too makes clear in the *Republic*' (5.593.3-4) The 'younger' authors on the other hand used the same name, viz. geometry, for both disciplines,[112] because both are concerned with the knowledge of magnitudes. So they connected them, converting them so to speak into a single study (πραγματεία), 'because, as has been said, they deal with the same thing'. This is an implicit description of the σκοπός of geometry in the later sense of the term: the subject of this discipline is magnitudes.

As we see, there is nothing about these introductory scholia which is particularly Neoplatonic.

---

[108] See also below, pp. 126-7, complementary note 108.
[109] δεῖ γὰρ ἐπὶ παντός, ὡς εἴρηται, πράγματος (isagogical terminus technicus) τὴν τῶν ἁπλῶν ἡγεῖσθαι διδασκαλίαν.
[110] For the rule in question see Mansfeld (1994) 112-3 n. 195.
[111] It is discussed together with *Schol. vat.* X.62 by Heath (1926) 3.1-3, whose treatment is slightly out of date, see n. 92 above.
[112] For Heron's use of the term geometry see below, text to n. 178; presumably he belongs with the 'younger' authors mentioned in *Schol. vat.* XI.1.

III 3 *Pappus' Commentary on* Elements *Book X*

We may now turn to Pappus' Commentary to *Elements* book X. Two preliminary issues have to be discussed first.

Jones argues that the Commentary may be the lost book I of Pappus' *Collectio*, basing this argument on entry 604 in a catalogue of the papal library at Viterbo written in 1311 by a librarian who, so he argues, knew no Greek. This begins with the words 'item unum librum, qui dicitur Commentum Papie super difficilibus Euclidis et super residuo geometriae, et librum de ingeniis'. 'Papie' must be Pappus. *Vaticanus graecus* 218 contains the remains of the *Collectio* and on its first page part of a work written in another hand (which also supplied some pages in Pappus), which Jones identifies as the Περὶ παραδόξων μηχανημάτων of Anthemius of Tralles. The formula 'librum de ingeniis' probably refers to this work. Ergo, thus Jones, the 'Commentum Papie super difficilibus Euclidis', or Commentary on *Elements* book X, is the lost book I of the *Collectio*. But in the first place this is not easily reconciled with the fact that the Commentary on book X is itself divided into two books. In the second place, the title 'Commentum ... super difficilibus Euclidis', which very much resembles a title of Heron transmitted in the *Fihrist*, viz. 'Book on solving the uncertainties of Euclid', perhaps refers to a separate work. This may or may not have been a part of Pappus' Commentary on Euclid. *Vat. gr.* 218 in its complete state may well have contained two different works by Pappus; the *Collectio* after all may have been copied from an already defective ancestor: since part of book VIII has gone missing, in its ancestor the beginning too may already have been lost already. The librarian's 'unum librum' is far less decisive than the explicit reference to the Commentary in *two* books on book X in the scholia on Euclid. Even if the ms. (or its ancestor, from which the description in the catalogue may derive) originally contained Pappus' Commentary on *Elements* X, it still does not follow that this originally was the beginning of the *Sunagôgè*.[113]

---

[113] See Jones (1986b), who disagrees with Grant (1971) 666-7, according to whom the formula 'librum de ingeniis' pertains to the abstracts from Heron's *Mechanica* at *Coll.* 8.31-2, and with Clagett (1978) 406 n. 56, who accepts Grant's view and argues that "whole entry" in the ms. "refers to Pappus' Collectio". I would add that it is equally possible that 'librum de ingeniis' pertains to the whole of *Coll.* VIII., and that 'unum librum, qui dicitur Commentum Papie super difficilibus Euclidis et super residuo geometriae'

The second issue pertains to the book's supposedly Neoplatonic colouring. Jones believes that the work "seems to have been composed for readers versed in philosophy, especially Neoplatonism ", and similar remarks are made by others. But there is no trace whatever of specifically Neoplatonic doctrines. A better interpretation is provided by Burkert, who writes: "in general his [viz. Pappus'] exposition is strongly influenced by Platonism ". In fact the Commentary on book X of the *Elements* is no more Neoplatonic than Nicomachus' *Introductio.*[114]

I go on with the text itself. Paragraphs of Pappus' text are quoted according to Thomson's translation, italics are mine. The first paragraph of book I begins as follows: 'The *aim* of Book X of Euclid's treatise on the elements is to investigate the commensurable and the incommensurable, the rational and irrational continuous quantities'. So Pappus begins with a description of what came to be called a book's σκοπός. A *historical excursus* follows; the origins of this theory, he tells us, are to be sought in the school of Pythagoras, but it was further developed especially by Plato's pupil Theaetetus, as Plato shows in the dialogue called after him, though later also the great Apollonius made important contributions. 'Eudemus the Peripatetic'[115] is cited for a description of Theaetetus' findings. 'Euclid's *object*, on the other hand [i.e. as different from that of Theaetetus], was the attainment of irrefragable principles which he established for commensurability and incommensurability in general'. In other words, what Pappus does here is justifying the *authorship* of Euclid, i.e. the correctness of the ἐπιγραφή, in a way

---

may indeed be a designation of *Coll.* II-VII, in which Euclid is one of the earliest authors (and, by reputation, the most important) to be treated. Jones' argument is criticized by Vanhamel (1989) 373-6, who reviews the literature on this issue and, perhaps wisely, opts for a *non liquet.* For Heron's title below, p. 126, complementary note 77; for the reference in the scholium below, n. 120.

[114] For Jones' hypothesis about the Commentary on book X see (1986a) 1.46-7, cf. Jones (1986b) 24-6. For its purported Neoplatonic ingredients see Jones (1986a) 1.11 (but cf. above, n. 98, and below, n. 119 and text thereto). For Burkert's more correct view see his (1972) 461 n. 68. "Some doubts " as to the authenticity of the Commentary are voiced by Bulmer-Thomas (1974) 293 and 299, who follows the obsolete Suter, cf. above, n. 98; he too speaks of the work's "Neoplatonic character ". For Pappus on Nicomachus see above, text to n. 68, n. 69, and below, Appendix 2. See further below, n. 121.

[115] This portion of the text is now reproduced as Fr. 141 I in the second ed. of Wehrli; see already Burkert (1972) 440-1 n. 82, 457-8 (quotation of part of Pappus §§ 1-2), 462 n. 73.

which is the same as that of *Schol.* V.3 and *Schol. vat.* V.1.[116] We may perhaps call this 'qualified *authenticity*': in its present *systematic* state the book is by Euclid, though it incorporates the work of his predecessors.[117] I have italicized the word 'systematic', since Pappus' remark at the same time pertains to the *ordering* (τάξις) of the contents.

Repeating the main issue of the previous paragraph at the beginning of § 2, Pappus goes on to deal with the χρήσιμον: 'Since this treatise has the aforesaid *aim* and *object*, it will not be unprofitable for us to consolidate the *good* it contains'. This good is explained at some length in §§ 2-3. Pappus again appeals to history, and at some length to philosophy. The familiar Pythagorean story that the person who first revealed that irrationals exists was drowned is allegorized in a Neopythagorean or Platonist way.[118] In the first place, Pappus argues, it is perhaps better not to make such irrationals public; and secondly the soul which finds out about these things by accident loses its bearings and wanders about in the stream of coming into existence and passing away, which lacks measurement.[119] Therefore 'the Pythagoreans and the

---

[116] See above, text to n. 108, and below, pp. 126-7, complementary note 108.

[117] Compare the way in which Apollonius of Perga in the introductory dedications of the various books of his *Conica* comments on the achievements of his predecessors (incorporated by him) as compared with his own additions and systematization; see below on book I, text before n. 126. Also see the proem of the mathematician Diocles (early 2nd cent. BCE) at Toomer (1976a) 34, of the rhetorician Aelius Theon (1st-2nd cent. CE) *Prog.* 59.14 ff. Spengel, and already the proem (1.1) of an anonymous physician, viz. [Hipp.] *De victu*, probably mid-4th cent. BCE. The same claim is made by Heron (often, cf. e.g. below, text to n. 170), by Ptolemy (cf. below, text to n. 231) and by Theon (below, text to n. 265).

[118] For the traditions concerning the various versions of this story see Burkert (1972) 455 ff.—esp. 458 with n. 58 on Iamblichus, who *VP* 246-7 cites no less than three versions the last of which, 132.20-3 Deubner, is that reported by Pappus—but his view (*ibid.* 461) that Pappus qualifies the story as a "legend " is questionable. Pappus (1 § 2) tells us that there was a 'saying' current in the school of Pythagoras about the man who perished by drowning after disclosing the knowledge of surds, 'which is most probably a parable by which they sought to express' etc. So Pappus provides an allegorical interpretation in philosophical terms of a 'saying' he believes to be genuinely Pythagorean; cf. below, p. 127, complementary note 119, for the formula ἴσως ἠνίττοντο. This approach is in no way different from the allegorical interpretations of the Pythagorean *akousmata* found in a number of authors (Anaximander the Younger, Aristotle, etc.), and the interpretation itself quite possibly is not original with Pappus.

[119] This passage as reflected in *Schol Vat.* X 1 was used by Sachs in her attempted rebuttal of Pappus' authorship, see above n. 98. For the Greek text

Athenian stranger' (reference to Pl. *Leg.* 819a) counseled prudence.
Plato's counsel should be heeded, and Euclid's 'wonderful *clarity*'
appreciated. The hazards are to some extent obviated by the fact that
the irrational pertains to geometry only, not to numbers, as is explained philosophically and at appropriate length in the sequel.[120]
And in geometry it can be neutralized in a scientific way.

   § 4 deals with the isagogical issue of the 'arrangement [i.e.
systematic ordering, τάξις] of ideas in Euclid's propositions', which is
explained at some length; this at the same time amounts to a
treatment of the *division* of the book into sections, or *parts*, as is clear
from the summary of §§ 1-4 at the end of § 4: 'The aim [σκοπός],
profit [χρήσιμον], and divisions [διαίρεσις εἰς μέρη] of this book
have now been presented in so far as is necessary'.

   §§ 5-23 deal at length with the study of irrationals from a mathe-
matico-philosophical point of view. I shall publish something
elsewhere on this section in which Pappus demonstrates his
familiarity with Plato and Aristotle.[121] So I conclude the present
brief overview of the first book of Pappus' work with §§ 24-36. At the
beginning of § 24 he states: 'let us begin again and describe its
parts'. At the end of § 4 Pappus had said that the division into parts
had by now been given insofar as necessary. In the concluding
paragraphs he presents a far more elaborate division into no less
than *thirteen parts* ('in the first part', 'in the second part', etc.) The
contents of each part are summarized, and it is furthermore clear
that the *ordering* of these sections is both *didactic* and *systematic*.

   From the above survey, mainly based on the Commentary on a
particular book, it will have become clear that Pappus in his
Elements Commentary is familiar with a good many isagogical
issues, that he is fully aware of their didactic relevance, and uses
them both explicitly and systematically. It is a pity that the

---

and some parallels see below, p. 127, complementary note 119.
   [120] It is perhaps to this paragraph and the next rather than to § 7 (*pace*
Jones [1986a] 1.10-1) that *Schol. vet. in Eucl. Data* nr. 4 refers (262.1-7 Menge *ad
finem*, cf. already Heiberg [1882] 163): 'both the rational and the irrational
can be a given [*datum*], as Pappus says at the beginning of (his Commentary)
on (book) X of Euclid ('s *Elements*)', δύναται δὲ καὶ ῥητὸν καὶ ἄλογον δεδομένον
εἶναι, ὡς λέγει Πάππος ἐν ἀρχῇ τοῦ εἰς τὸ ι΄ Εὐκλείδου· τὸ μὲν γὰρ ῥητὸν καὶ δεδομένον
ἐστίν, οὐ πάντως δὲ καὶ τὸ δεδομένον ῥητόν ἐστιν.
   [121] For philosophy in the *Sunagôgè* see below, Appendix 2, and the *haute
vulgarisation* version at Mansfeld (1998a). A paper on the philosophy in the
Commentary will appear elsewhere. There are important links with the
philosophy in the *Sunagôgè*.

introductory part of his Commentary dealing with the *Elements* as a whole is no longer extant, for one would have liked to know what his presentation of the author and his treatise could have resembled. Perhaps he used Eudemus' *History of Geometry,* just as at the beginning of the part that has been preserved. Even if the Commentary on book X was composed first (which to some extent would explain its lengthy treatment of a number of isagogical questions), that to book I and the treatise as a whole can hardly have been less rich. Whether some of the issues dealt with in Proclus' Commentary on book I derive at least in part from Pappus' Commentary is a matter for speculation. To answer the question whether Proclus knew and used the Commentary on book X more research is needed.

CHAPTER FOUR

APOLLONIUS' PROEMS AND EUTOCIUS' COMMENTARY

IV 1 *The Proems of Apollonius'* Conica

Four of the eight books of Apollonius of Perga's *Conica* are extant in
Greek, together with a Commentary by Eutocius of Ascalon.[122]
Apollonius is a great mathematician, admired but also criticized
by Pappus, who has also preserved information about the books of
the *Conica* lost in Greek and about other lost works, both in the
Collectio and in the *Commentary on Elements* X.[123] The final version of
the *Conica* (in instalments) presumably has to be dated not too long
after 200 BCE.

   Of great interest in our present context are Apollonius' proems to
the individual books; these are in the form of letters to the
dedicatees: Eudemus, the first teacher of the Epicurean philosopher
Philonides,[124] for books I-III, a certain Attalus for books IV-VII (and
VIII, I presume) after Eudemus' death.[125]

---

   [122] Ed. Heiberg (1891-3), including Eutocius' Commentary (for which see
Ch. IV 2). Books V-VII are extant in Arabic (book VIII being lost), and are
now acccessible in Toomer (1990) which replaces Halleius (1710); note that
Toomer's remark at (1990) 1.vii that Halleius failed to print the Arabic text is
a slip. The *Conica* belongs with the domain of Analysis, see above Ch. II. On
their mathematical contents see Heath (1921) 2.154-75, Toomer (1970) 181-8,
and Toomer (1990) 1.xiv-v and xxviii-xxxiv esp. for books V-VII. For Apollonius' dates see Toomer (1970) 179-80 and (1990) 1.xi-xii: his son and messen-
ger was an adult, and Philonides is allowed to see the work (proem to book
II; see below).
   [123] Reprinted from Hultsch (1876-8)—including the mathematical lemmas on the extant books—and Woepke (1856) at Heiberg (1891-3) 2.102-66,
together with fragments cited from Eutocius' Commentaries on Archimedes,
from Philoponus, Proclus, Hypsicles (i.e. *Elem.* XIV), Marinus, Ptolemy,
Hippolytus, Ptolemaeus Chennus, and the *Fragmentum Bobiense*. The section
derived from Woepke (1856) at 2.120-4 Heiberg should be corrected on the
basis of Thomson (1930).
   [124] *Pap. Herc.* 1044 Fr. 25.4-5, see Gallo (1980) 33 and 36.
   [125] The proems to books I-II and IV-VII are translated and discussed by
Heath (1896) lviii-lxxxvi, i.e. those to books I-II and IV are translated from
Heiberg's Greek text, that to book V from Nix's Latin (1889), and those to
books VI-VII from Halleius' Latin (1710). I have consulted Heath's transl. for
books II and IV, that of Toomer (1990) for the proems to books V-VIII, as well
as Toomer's new translation of the proem to book I at (1990) 1.xiv-xv. On the

In the introduction to book I (1.2-4 Heiberg) he writes to Eudemus that he sends him the revised version of this book, and that the others will follow as soon as they have been revised too. Drafts of books I-VIII already exist: the work was written at the request of the geometer Naucrates when this colleague was staying with Apollonius at Alexandria, and Apollonius (or so he claims) hurriedly (!) jotted down a preliminary version of the whole treatise in eight books and gave this to his friend, who had to leave Alexandria. This remark about an earlier dedicatee (?) and to hurried composition sounds a bit like a *topos*, but this is by the way. Copies of this preliminary version of books I and II had since also been given to other friends. Eudemus should therefore not be surprised when encountering versions different from the present corrected and polished, i.e. an authorized edition. The preliminary version therefore cannot have been very rudimentary. Revision must have been a matter of style, of adding prefaces, etc.

Apollonius then meticulously informs Eudemus (and so the general public) beforehand about the *contents* of the whole treatise. Books I to IV deal with the elementary instruction; next, the *contents* of *each book* are announced and summarized (περιέχει ... τὸ πρῶτον [*scil.*, βιβλίον], ... τὸ δεύτερον, etc.) The first book deals with matters that have been already treated by others (no names given), but according to the author it does so in a fuller and more general i.e. *systematic* way. Nevertheless, what we have here is a reference to the *history* of the subject. The specific *utility* (γενικὴν καὶ ἀναγκαίαν χρείαν) of the contents of book II is emphasized. Book III contains a great number of theorems which are *useful* (χρήσιμα) for the synthesis of solid loci etc.[126] Most of these are new, that is to say have been found by Apollonius himself, or so he claims. Greek mathematicians are not averse to the idea of progress! Euclid's treatment of a specific issue, for instance, is said to be both incomplete and unsystematic—an affirmation which produced an interesting controversy.[127] The contents of book IV, he tells us, are for the most part original.

---

final section of the proem to book I see Friderici (1911) 43-4.

[126] Cf. above, n. 26 and text thereto; below, p. 123, complementary note 26.

[127] Cf. above n. 19, below text to n. 131, and n. 139 and text thereto. Toomer (1970) 180 and 186-7 argues that Apollonius in books I-IV for the most part systematized the findings of his predecessors, among whom Archimedes (whom he fails to mention by name in the preface to book I). So this part of his work would be of the same nature as most of Euclid's *Elements*.

The other books, Apollonius says, go much further than the elementary and general instruction provided by books I-IV; in the briefest terms he tells us what is the subject of each of them. Book V is about maxima and minima, book VI about equal and similar conic sections, book VII about theorems concerning diorisms, and book VIII about determinate conic problems.

We may notice that isagogical questions dealt with systematically in the literature of later and late antiquity are already present in a preliminary way in the general introduction to the first book: the *theme* of the work as a whole and the subjects of the individual books (entailing in some cases *historical* references, viz. remarks about predecessors in the same field, one name even being mentioned), the specific *utility* of some of its parts, the *division* of the work into *parts* and *subparts*, i.e. *two* main sections consisting of *four* books each, and the *systematic order* of these two main sections and of the individual books which coincides with the *order of study*. We must further note the justification of this revised edition itself and the reference to the earlier draft versions, that is to say the distinction between draft versions which may circulate among colleagues and pupils and have been copied by others, and the official edition as corrected by the author. This topic is often an issue in the introductory sections of for instance Galen as well, about five centuries later.[128] The combination, in this brief compass, of a justification of the corrected edition from a literary and historical point of view with a survey of its contents is to some extent comparable with Porphyry's justification, in the *Vita Plotini*, of his corrected edition of Plotinus' works in an ascending systematic order, with titles revealing their specific themes.[129]

It is worthwhile to compare Pappus' remarks in his introduction

---

[128] Apollonius' account is Devreesse's earliest example for this practice (cf. above, n. 17 and text thereto). For the working methods of ancient authors see Dorandi (1991b). Attalus of Rhodes, who according to his proem quoted by Hipparchus *in Arat.* 1.1.3 = Attalus Fr. 1, 3.11-20 Maass (1898) published an *editio correctior* of someone else's work, viz. Aratus (see Mansfeld [1994] 162 with n. 295) is probably to be dated to ca. 150 CE, see e.g. Kidd (1997) 18; that he is to be identified with the dedicatee of *Con.* IV-VIII can be no more than speculation (Toomer [1990] 1.xii n. 2). Attalus writes to his unknown dedicatee that he has *sent* the book of Aratus which he has *corrected* (τὸ ... τοῦ Ἀράτου βιβλίον ἐξαπεστάλκαμέν σοι διορθωμένον ὑφ' ἡμῶν, and a little later: τὴν διόρθωσιν τοῦ βιβλίου), plus his interpretation (ἐξηγήσιν) which makes Aratus' views agree with the phenomena.

[129] Cf. Mansfeld (1994) 108-16. Also think of Galen's autobibliographies discussed *ibid.* 117-31, or of Possidius' *Vita Augustini.*

to the discussion of the *Conica*, 2.672.30 ff. Hultsch. These are heavily dependent on Apollonius' dedication/introduction to book I, a substantial chunk of which (viz. 1.4.1-26 Heiberg which as we have seen summarizes the contents of the work as a whole) is even quoted practically verbatim at 2.674.22-676.18. Pappus' designation of this passage is interesting: 'Apollonius says what the eight books of *Conica* written by him contain, placing a preliminary heading-like clarification in the proem of book I', 2.674.20-1. Interesting not only because Pappus correctly calls the dedication/introduction a 'proem', but also because he calls this summary of the contents a κεφαλαιώδη προδήλωσιν, a 'preliminary heading-like *clarification*', i.e. one listing in a clear way the main *themes*. The substantive προδήλωσις ('announcement', 'prediction', cf. the meaning the verb ususally has) is very rare—in the *Collectio* it occurs only here—, and its present meaning is not listed in the new LSJ. The formula as a whole is an excellent designation of what an introduction should contain in the matter of a listing of topics. Furthermore, Pappus, a partizan of Euclid, argues that Euclid's *Conica* in four books[130] were merely 'filled out' by Apollonius (2.672.18), and he defends the Elementarist against what he believes to be Apollonius' unjustified criticism (2.676.19-8.12).[131]

But let us return to Apollonius himself. The proem to the next book is brief to a degree (1.192.1-11): he merely says that his son is now bringing book II, recommends that it be studied carefully and permits that it be communicated to those who deserve this, Philonides being mentioned in particular. Book III has no proem, so presumably the authorized version has been lost.[132] That to book IV (2.2-4) on the other hand is quite substantial. Apollonius writes to

---

[130] I cannot enter into the problem of the existence of this work; for the issue see Jones (1986a) 2.399-401.

[131] Cf. above, text to n. 127. For Eutocius on Apollonius' originality see below, Ch. IV 2.

[132] Eutocius 2.314.4-5 Heiberg tells us that book III lacks a dedication, and 2.354.6-7 that he has 'edited' book IV; at 176.17-20 he tells us that he has edited all the books from the various copies available to him. Since Apollonius in the proem to book IV advises us that the *three* previous books had been dedicated to Eudemus, and books IV-VII are dedicated to Attalus, the authorized version of book III sent to Eudemus and including the dedication was no longer available to Eutocius. Eudemus' death may have been the reason why the final version of book III did not circulate widely enough. The alternative hypothesis, viz. that the proems of the *Conics* are spurious additions, is avoidable; see above, n. 128, below, n. 238. Their authenticity has never been questioned.

Attalus that he has given books I to III of his *Conica* to Eudemus, but that beginning with book IV he will dedicate them to him as his new dedicatee, for Eudemus is now dead.

So here is book IV. Its *contents* are listed (περιέχει δὲ τοῦτο κτλ.), which fall into *three sections* (2.2.13 περὶ τοῦ δευτέρου, 2.22 τὸ μέντοι τρίτον). The description of these sections includes a short *historical* overview: what belongs with the first section has been treated by Conon of Samos in his *To Thrasydaeus*, but incorrectly. Nicoteles of Cyrene then wrote against Conon, but as to what belongs with the second section he only indicated that proofs could be given but failed to do so himself; neither did anyone else. Finally, what is in the third section has never been treated before. The new theorems in books I-IV are said to be very *useful* for what we may call 'higher' conics.

The proem to book V is quite substantial; those to books VI and VII are shorter. On the whole the descriptions of their *contents* are similar to those in the books extant in Greek. In that to book V he writes to Attalus that his predecessors have hardly paid attention to the theory of minima lines. Insofar as they have come near this topic their views have been incorporated in book I, but apposite treatment and proofs concerning minima will be provided only now, with treatment of maxima and several related issues thrown in. We again note the careful distinction made between his own achievements and those of his predecessors. In a similar way book VI is to treat matters which have been neglected by Apollonius' predecessors, at least in the sense that his treatment will be both richer and *clearer; inter alia* conic sections which are equal to each other or dissimilar to each other, as well as segments of conic sections will be dealt with. Book VII too contains a number of new theorems, which are of great *use* for many types of problems. They will also prove *useful* for solving problems to be discussed in book VIII, which is to follow.

IV 2 *Eutocius' Commentary on Apollonius'* Conica *I-IV*

Turning now to Eutocius' *Commentary* on books I-IV (which is later than the Commentaries on Archimedes to be discussed in the next chapter),[133] we must note that the Commentary on book I is quite

---

[133] On Eutocius see Heiberg (1880), Bulmer-Thomas (1971), Knorr (1989)

substantial and has a quite long introduction (which in fact intro-
duces the whole work), while the commentaries on the other
books are much shorter (II-III) to extremely short (IV), the prologues to books II-III being very brief, that to book IV again longer.

In his introduction to the Commentary on book I Eutocius
(2.168.5-186.21 Heiberg) first attempts to give Apollonius' relative
date, citing the *Life of Archimedes*[134] written by Heraclius. This
Heraclius (or Heraclides)[135] argued that the conic theorems had
been discovered by Archimedes but had not been published by
him, and that Apollonius had appropriated them. We may view
this quote (concerned with the life as well as with the work) as an
echo of sorts of the *Vita* which may stand in front of the edition of
the *first work* of an author *to be studied*,[136] Eutocius only using what
meets his particular purpose. For according to him Heraclius'
claim is mistaken. In the first place, he states, Archimedes often
appears to allude to the Στοιχείωσις τῶν κονικῶν, i.e. Apollonius'
Conica, thus showing that it is earlier than his own work.[137] The
term στοιχείωσις, recalling Euclid's title, shows that according to
Eutocius Apollonius' treatise was the fundamental work on the
subject. Secondly, Apollonius does not pretend that he expounds
his own discoveries alone, for he tells us that he has treated in a
fuller and more systematic way matters that had been already
written about by others. Eutocius next paraphrases a discussion to be
found in book VI of Geminus,[138] who (in his view correctly)
pointed out the difference between the systematic and general
account of Apollonius and the efforts of his predecessors. Pappus'
discussion of the same issue in the *Collectio*, misreported here
(2.186.1-10), was known only indirectly to Eutocius.[139] But the fact
that both these men do discuss Apollonius' explicit criticism of

---

225-6, 229-31, 233-8, Toomer (1990) 1.xvi-xvii, Decorps-Foulquier (1997). For his
date see below, n. 143.
[134] See below, Ch. V 3.
[135] This is his name at Eutoc. *in Arch. De dim. circ.* 3.228.20-1 Heiberg,
where the *Vita* is also cited.
[136] It may also be found at the beginning of the Commentary on such a
work, as in Olympiodorus' Commentary on the *Alcibiades maior*: the γένος τοῦ
φιλοσόφου at *in Alc.* 2.14-167 Westerink. See further Mansfeld (1994) 179-91.
[137] False, since Archimedes died in 212 BCE, i.e. possibly even before
Apollonius began drafting his *Conica*, and certainly before he began
publishing it.
[138] For Geminus see above, n. 81 and text thereto.
[139] Above, n. 19. For Pappus and Eutocius on Apollonius in relation to his
predecessors cf. Fraser (1972) 1.428-32.

Euclid, albeit in various ways, shows that this debate, which in fact goes back to some extent as least as far as Geminus and Heraclius/ Heraclides, went on for centuries.

Comparison with remarks found in Pappus' Commentary on Euclid *Elements* X, in the *Collectio*, and in the *Scholia vaticana* on Euclid, further shows that one of the issues involved in this introduction is the *theme* of the work in relation to the question of *authenticity*; that is to say the relation of Apollonius to predecessors in the same field is comparable to that of Euclid to his predecessors, e.g. Eudoxus and Theaetetus.[140] Eutocius also discusses the *contents* of the individual books at some length (books I-III at 2.176.23-80.10, book IV at 2.186.11-21), mostly cannibalizing Apollonius' own introductory 'epistle' (as he calls it, 2.176.23) to book I, but adding interesting comments. He follows his source as to the *division into parts*. This preliminary account allows the prologues to the following books to be as short as they are.

Furthermore, Eutocius informs us that in the text of his edition he has put together the *clearer* parts to be found in the different versions at his disposal,[141] for the benefit of *beginners* (τὰ σαφέστερα παρατιθέμενος ἐν τῷ ῥητῷ διὰ τὴν τῶν εἰσαγομένων εὐμαρείαν), while his own comments and passages he feels bound to exclude are written in the margins (2.176.17-22). So half-way the long proem *clarity* too is mentioned, *disertis verbis* this time, while the *qualities to be expected of the students* also play a role.

In the short prologue of the Commentary on book II (2.290.1-5) Eutocius states that he will only write about those things which cannot be understood on the basis of what he has written on book I. The proem to book III (2.314.1-11) is a bit longer. Eutocius tells us that this book was much esteemed by the ancients, as is made clear by the existence of various versions. Still, it lacks an introductory letter (i.e. dedication), and no σχόλια worth anything written by those 'before us' (πρὸ ἡμῶν) are to be found, though Apollonius in the proem to the whole work says that the contents of book III are well worth looking into. Eutocius' own *clear* explanation (σαφῶς ... δεικνύμενα), based on (what is in) the previous books and his comments on these books, is now available to the *student*. The proem to book IV (2.354.1-356.4) briefly lists its contents, praises its

---

[140] See below, pp. 126-7, complementary note 108.
[141] For Eutocius' methods see above, nn. 27 and 39, Knorr (1989) 237-8, and Decorps-Foulquier (1997).

clarity for those who read it, especially in his, Eutocius', edition, and confesses that it does not lack [earlier] σχόλια, for what is lacking (viz. in Apollonius' exposition) is filled out by what is written *in margine* (αἱ παραγραφαί).[142] The method of proof throughout the book is by *reductio ad absurdum*, also used by Euclid, Aristotle, and Archimedes. 'Not lacking in [earlier] σχόλια': as a matter of fact Eutocius' own comments on book IV barely fill three Teubner pages, and the abundant earlier 'scholia' are lost ... Eutocius survived, his predecessors did not. Anyhow, Eutocius continues, if one *studies* (ἀναγινώσκοντι) books I-IV one will be in a position to solve problems in the field of conics, for these books contain all one needs by way of elementary information, the remaining books, as Apollonius himself has said, being a lot more specialized. So diligent study of books I-IV plus Commentary is recommended (ἀνάγνωθι οὖν ταῦτα ἐπιμελῶς), and if the reader wants Eutocius to expound the other books in the same way this, God willing, will be done. Presumably it never was.

One notes that in the Apollonius Commentary Eutocius is not interested in isagogical questions in a systematic way though quite a few are unmistakeably present; this surprises one a bit since he is a pupil of the Neoplatonist philosopher and commentator Ammonius Hermiae;[143] Ammonius, as is well known, liked and used the rigid isagogical schemes.[144] But see the next chapter, on the earlier Commentaries.

---

[142] We know something about Eutocius' predecessors. According to the *Suda* lemma on Hypatia (Y 166, 4.644.4-5 Adler) this lady philosopher wrote a Commentary on the *Conica* (lost). Serenus tells us he wrote a Commentary on the (first book of the) *Conica* (lost as well), 52.24-7 Heiberg: ὡς ἐν τοῖς Κωνικοῖς [Apoll. 1.15] δείκνυται ... καὶ ἡμεῖς ἐν τοῖς εἰς αὐτὰ ὑπομνήνασι γεωμετρικῶς ἀπεδείξαμεν (cf. above, n. 8).

[143] For Eutocius' date and relation to Ammonius see Knorr (1989) 229-30. He may have presided over the school at Alexandria after the master's death and before the succession by Olympiodorus. On this period see Verrycken (1994) 44-8, with references to the literature.

[144] Cf. above, n. 10.

CHAPTER FIVE

EUTOCIUS' COMMENTARIES ON ARCHIMEDES,
AND THE *VITA*

V 1 *Archimedes' Proems*

The introductory letters/dedications of Archimedes' works (note
that some are extant without such an introduction) tell the reader
quite a bit in a traditional way about the *content* of the treatises involved and the occasions which prompted him to write and send
them. But unlike Apollonius he apparently is not interested in isagogical issues as such. This difference with Apollonius is perhaps
capable of being explained. I would suggest that Apollonius, living
and working at least for some time in Alexandria,[145] had been
influenced by the methods of Alexandrian philology, that is to say
the editing and publishing of corrected standard versions of the
great classical authors. Think of his careful distinction between
drafts, or various versions, on the one hand and the polished and
authorized ἔκδοσις meant for the general public on the other.
Archimedes, for his part, though maintaining a lively exchange
with the mathematicians of Alexandria, as appears from several of
his dedications,[146] lived and worked in far-away Doric-speaking
Syracuse, and shows no interest in new-fangled modes of
presentation.

---

[145] Though he moved around (we know from his proems that he had
visited colleagues at Pergamum and Ephesus) he lived long enough in
Alexandria to compose the first draft of the eight books of *Conica*, and he
already was a resident of the city when the colleague for whom he wrote it
came and stayed with him. The revised versions are *sent* to the dedicatees at
Pergamum; so from elsewhere, most probably from Alexandria. The proem
to the authorized version of book I at least suggests that the author was still
living in Alexandria at the time.
[146] See e.g. Fraser (1972) 1.399–402.

V 2 *Eutocius' Commentaries on Archimedes*

The earliest[147] of Eutocius' Commentaries,[148] that on the first book of Archimedes' *De sphaera et cylindro* has a short prologue in the form of a quite flattering dedication to his 'master' Ammonius (3.2.1-22 Heiberg). Note moreover that Eutocius also includes the first part of his account of the definitions in the 'introduction' (ἐν τοῖς προοιμίοις τοῦ Περὶ σφαίρας καὶ κυλίνδρου, *In plan. aeq.* 3.268.14-5). Eutocius states that his motive for attempting to write on this difficult treatise, which absolutely needs to be explained, is that there is a gap: 'I found that no one before us has composed a worthwhile work' (οὐδένα τῶν πρὸ ἡμῶν ἀξίαν εὑρὼν σύνταξιν καταβεβλημένον), viz. dealing with this book. He repeats this: no one before has approached this *subject* (ὑπόθεσιν; one is pleased to encounter an isagogical terminus technicus). Another isagogical issue is of course also present, viz. the *difficulty* of the subject which needs *clarification* (σαφῶς ἐκθέσθαι τὰ ... δυσθεώρητα).[149] Also, the very first sentence of the introduction cites the ἐπιγραφή (τὰ Περὶ σφαίρας καὶ κυλίνδρου Ἀρχιμήδους), about the *authenticity* of which there clearly is no doubt and which need not be further explained. It is interesting to note that Eutocius *ad finem* uses σκοπός in the sense of (his, Eutocius') authorial intention, though in a semi-proverbial expression. He asks Ammonius to tell him what he thinks of the work (γράμμα); 'if it has not altogether missed its aim' (εἰ δὲ τοῦ σκοποῦ μὴ πάντη διαμαρτάνον), its author will try to write on other Archimedean works as well.

Clearly the Commentary on book I did find favour with the venerated Ammonius, for in the very short prologue to book II (3.50.2-4) Eutocius declares that, having *clarified* (σαφῶς ... γεγραμμένων) the

---

[147] Eutocius apologizes for possible mistakes due to his youth (3.2.12-3, εἴ τι παρὰ μέλος διὰ νεότητα φθέγξομαι).

[148] Ed. Heiberg (1910-5) vol. 3, Mugler (1972) with French transl. The Commentaries on the *De sphaera et cylindro* and the *De dimensione circuli* have been 'edited and collated' (ἐκδόσεως παραναγνωσθείσης) in antiquity by Isidorus of Miletus, as end-notes tells us. Presumably they started their career as text-books (presumably collected in codices) for a small circle of users. Two Commentaries on Archimedes by Eutocius (*Eutokii Ascalonite rememoracio in libros Archimedis de spera et chylindro* and *Euthocii Ascalonite rememoracio in libros Archymedis de equerepentibus*) and seven treatises by Archimedes have been translated by Willem van Moerbeke, see Vanhamel (1989) 362-7.

[149] Clarity and clarification are often at issue, see Heiberg's index ii, 3.437, *s.vv.* σαφήνεια, σαφηνίζω, σαφής.

theorems of the first book, he will now perform the same service for those of the second. In this second book various interesting items are to be found. At the beginning of the abstract from book VIII of the *Collectio*[150] he speaks of Pappus' πρόθεσις, his *authorial intentention* (3.70.9, cf. 7, προέθετο). At 3.150.13 (cf. 152.14) he speaks of the πρόθεσις of Archimedes in the *De sphaera et cylindro*. At 3.132.5-18 he is concerned with the *clarification* (σαφεστέρῳ ... λέξει γράφομεν) by translating his difficult Doric and replacing his archaic terminology. As to Nicomedes' *De conchidibus*, he says that this *title* was given by the author himself (3.92.2-3, Νικομήδης ἐν τῷ ἐπιγεγραμμένῳ πρὸς αὐτοῦ Περὶ κογχοειδῶν συγγράμματι). So no doubt about the work's *authenticity*.

The prologue (3.228) to the Commentary on Archimedes' opusculum *De dimensione circuli*[151] is from our point of view also quite rewarding. Eutocius states that he will 'achieve his aim' (σκοπός) by explaining those passages in Archimedes which need to be explained (briefly if they are relatively *clear*, others more fully) by linking up these explanations with his Commentary on the *De sphaera et cylindro*. The next text to be treated is τὸ γεγραμμένον Ἀρχιμήδει βιβλίδιον Κύκλου μέτρησιν τὴν ἐπιγραφὴν ἔχον, ἐν ᾧ τὴν πρόθεσιν τἀνδρὸς ἐξ αὐτῆς τῆς ἐπιγραφῆς γνωρίζομεν, 'the little book written by Archimedes which has as its *title* <<Measuring of the Circle>>, in which we learn the *author's intention* from the title itself'. What this title means is explained in the following colon. This, beyond doubt, is a conscious use of three preliminary issues, though not in the usual order; *authorship, explanation of the title, authorial intention*. Eutocius includes a short *historical overview*, which is most apt whenever authorial intention is the issue; he refers to the efforts of Hippocrates of Chios and Antiphon which, as he supposes, will be familiar to students of Eudemus' *History of Geometry* and Aristotle's writings (*scil.*, *SE* ch. 11). A reference to Heraclides in the *Vita*[152] follows, who had said that this little book is 'necessary for the uses of life'[153] and so, we may add, already dealt with the issue of the χρήσιμον; Eutocius accepts this interpretation.

---

[150] See above, text to n. 18.
[151] On the transmission and interpretation of this tract in antiquity and the middle ages see the account of Knorr (1989) 375-816, a book within a book.
[152] See above, text to nn. 134 and 135.
[153] Cf. above, n. 71.

This Commentary also contains an interesting afterword (3.258. 15-60.9). Eutocius admits that Apollonius of Perga's computation in the *Ocytocius* is more precise, but submits that this precision is not *useful* for Archimedes' *aim* (οὐ χρήσιμον ... πρὸς τὸν Ἀρχιμήδους σκοπόν). He refers back to his proem (3.128.19 ff.), where he has said—in fact by quoting Heraclides[154]—that this σκοπός is concerned with practical utility (διὰ τὰς ἐν τῷ βίῳ χρείας). The criticism of Sporus of Nicaea and other later authors is mistaken, since all have ignored Archimedes' σκοπός.

The prologue to the Commentary on book I of *De planorum aequilibriis* is rather short (3.264.2-15). It first refers to Aristotle, and to Ptolemy who follows him, for a definition of ῥοπή, i.e. the 'inclination of the scale', as the 'common genus of heaviness and lightness', then to 'Timaeus in Plato'. Those who are interested in the tenets of these authorities may collect them (ἔξεστι τὰς δόξας τοῖς φιλομαθέσιν ἀναλέγεσθαι κτλ.) from Ptolemy's *De momentis* (Περὶ ῥοπῶν),[155] from Aristotle's physical treatises, from Plato's *Timaeus*, and from those who have written Commentaries on these works. This advice is absolutely fascinating, at least to the present writer, since as a matter of fact Eutocius advises us that one may, or even should, compile one's own doxography,[156] and tells us how one should set about this.[157] Finally, he states what is the view of 'Archimedes in this book', thus implying that it is *authentic* and telling us what is its *aim*. The prologue to book II (3.278.1-3) is as short as can be and as to contents very much resembles its counterpart, the prologue to book II of the *in De sphaera et cylindro*.

We may sum up this overview of Eutocius' practice in the previous section and the present one by concluding that, though certainly familiar with the scholastic scheme of isagogical questions, and fully aware of the technical terminology involved his use thereof is quite unpedantic. This holds in particular for the

---

[154] Cf. above, text to n. 152.

[155] Lost; see Heath (1921) 2.295. The Eutocius text (incomplete) is Ptol. Fr. 3 Heiberg; Ptolemy's treatise is also cited by Simpl. *in Phys.* 710.14 ff. Diels (= Ptol. Fr. 1 Heiberg).

[156] There is no chapter Περὶ ῥοπῆς in Aëtius' *Placita*, though tenets of Plato and Aristotle on heavy, light, and ῥοπή (the latter in the Plato lemma only) are among the items treated *Plac.* 1.12 Diels, the chapter Περὶ σωμάτων. Perhaps Eutocius knew the *Placita* and was aware of what he saw as a lacuna.

[157] Possibly, Eutocius does not include this doxography because this would be a transgression of the boundaries of the genre (a mathematical Commentary). For a similar attitude in Proclus see below, text to n. 383.

longer prologues, where he attempts to write real literary prose. Often enough we have to infer that an isagogical question is at issue, and no instance can be given where all of them are present in some way or other at the same time.

From his references to the *Timaeus* of Plato, to the physical treatises and the *Sophistici Elenchi* of Aristotle, and to the Commentaries on the *Timaeus* and the physical treatises of Aristotle, it is clear that, though specializing in mathematics, Eutocius had received a solid philosophical education in the school of Ammonius.

V 3 *The* Vita *of Archimedes*

Finally, a word about the *Vita Archimedis* by Heraclius/Heraclides quoted several times by Eutocius.[158] As we have seen above this dealt both with the life in the proper sense of the word, and with the works. It provided a date for Archimedes, discussed questions of priority regarding some of his works in relation to Apollonius (the latter providing a *t.p.q.* for Heraclius/Heraclides),[159] and presented a view as to what measuring the circle is useful for. The obvious place of a *Vita* of this kind is at the beginning of an edition of the *opera omnia*,[160] but to the best of my knowledge we do not have any further information on whether such an edition existed.

---

[158] See above, text to nn. 134 and 152, where the passages involved are cited. The few facts we know about Archimedes' life as well as the anecdotes concerning him are discussed at Dijksterhuis (1956) 9-32; further literature at Knorr (1987) 421-2.

[159] Heiberg's guess (1910-5) 3.447 *s.v.* Ἡρακλείδας that he may be the Heracleides twice mentioned by Archimedes in the introduction to the *De lineis spiralibus* ("an idem? "), accepted by Fraser (1972) 2.600 n. 316, is not at all likely on chronological grounds.

[160] See Mansfeld (1994) 179-91.

CHAPTER SIX

HERON OF ALEXANDRIA

VI 1 *Introduction*

What survives of the voluminous works of Heron of Alexandria (to be dated to the mid-1st cent., as he mentions a lunar eclipse he observed in 62 CE)[161] is a rather mixed bag. For the most part these works pertain to applied mathematics and, again for the most part, they have not reached us in their original form, but underwent various revisions.[162] The fragmentary remains of his Commentary, or comments, on Euclid's *Elements* have been mentioned above.[163] In the present chapter I shall discuss the relevant sections of a number of works of Heron in the rather erratic order in which they are printed in the *Teubneriana*,[164] but begin with the *Belopoiica* which has been edited separately.[165]

VI 2 *The* Belopoiica

The first chapter of the proem (chs. 1-2) of this treatise on artillery is a shade bizarre. Heron first says that the 'most important and most indispensable part of philosophical study is that which is concerned with tranquillity of mind' (τῆς ἐν φιλοσοφίᾳ διατριβῆς τὸ μέγιστον καὶ ἀναγκαιότατον μέρος ὑπάρχει τὸ περὶ ἀταραξίας).[166]

---

[161] See Drachmann (1972) 310 and Neugebauer (1975) 2.846, referring to Neugebauer (1938) 21-4.

[162] E.g. Heath (1921) 307-10; Heiberg (1925) 37: "die echten Metrika [first published 1893 from a ms. in Constantinople] beweisen, daß die ... Geometrica, Geodaesia, Stereometrica und Μετρήσεις späte Rechenbücher sind, in byzantinischer Zeit in verschiedenen Redaktionen zusammengestellt".

[163] Text to n. 77; see further below, pp. 126-7, complementary note 77.

[164] Ed. in 5 vols.: Schmidt (1899), Nix and Schmidt (1900), Schoene (1903), Heiberg (1912-4). I shall not dicuss works which offer no information that is relevant in our present context: the *Mechanica* (which, "as preserved in the Arabic, is far from having kept its original form, especially in Book I ", Heath [1921] 2.346), and the *Stereometrica* and *De mensuris* (cf. above, n. 162).

[165] Diels and Schramm (1918).

[166] Diels (1893)107 = 245 calls this "sarkastisch", Heiberg (1925) 37 "ziemlich albern". Keimpe Algra points out to me that Heron stands the doctrine

Since we are now able to date Heron quite early, this reference to a Hellenistic *summum bonum* is no longer surprising.[167] Heron continues by pointing out that the philosophers have devoted—and still devote—the majority of their investigations to this issue, and believes their discussion will never end (a clear hint at disagreement, διαφωνία). But he has a solution: mechanics has left these theoretical discussions by the wayside, and taught all men to attain tranquillity with the help of a single and very small part of itself, viz. the science of artillery. One need not be worried about attacks, either from outside or from inside. So *Belopoiica* has to be studied and practised at all times. An interesting way to tell us that this discipline is subsumed under mechanics (the isagogical issue ὑπὸ ποῖον μέρος ... ἀνάγεται). In the next chapter Heron states that his predecessors have failed to deal in the proper way with the construction and use of the machines that are involved; this is what he intends to do in a manner which all readers will be able to understand (issues of *clarity* and of the *qualities of the students*). He then describes the *orderly and systematic* way in which he will treat these matters, both generally and in detail.

VI 3 *The* Pneumatica

Of the two books of which the *Pneumatica* consists only the first section (p. 4-10 Schmidt) of the long introduction (p. 4-28.15) to book I need be looked at here. In his backward reference to this introduction Heron uses a term which belongs with the later isagogical terminus technicus προθεωρία, viz. the verbal form προτεθεωρημένων (p. 28.17). We may limit ourselves to this section (certainly by Heron himself), because the extensive second section, however indispensable for what follows, is a justly famous philosophical argument concerned with the void deriving (at least to some extent) from the Peripatetic scholarch Straton of Lampsacus.[168]

---

of Epic. *Sent.* 6 and 7 (*ap.* D. L. 10.140-1) on its head.

[167] Diels (1893) 107 = 245 n. 1 says "Ich halte freilich auch diese Einleitung für compilirt aus älterer Quelle", presumably because he did not exclude a later date for Heron (cf. *ibid.* 106 = 244 with n. 6: "frühestens am Anfang unseres Zeitalters"). In the 1st cent. CE the main Hellenistic schools were still very much alive.

[168] Frs. 56, 57, 64, 65b, 66, 67 Wehrli. See Diels (1893), Drachmann (1948) 90-2, Gottschalk (1965). For the way in which Heron attempted to confirm this theory by experiments see Crombie (1994) 1.179-81.

In his proem Heron states that the ancient philosophers and practitioners of mechanics have payed much attention to pneumatics, some concentrating on its theoretical, others on its visible aspects. This is an implicit reference to the mathematical *subdiscipline* pneumatics *belongs with*, viz. mechanics.[169] He views it as his duty to bring into *order* (εἰς τάξιν ἀγαγεῖν) what the ancients have transmitted, and to add, or insert, what he has discovered himself.[170] This will be most *useful* (ὠφελεῖσθαι) for future mathematicians. We notice that in this way the *aim* of the treatise is made clear too, though only implicitly. The present work is the sequel to an earlier one in four books dealing with water clocks (lost)—so an *order of study* seems to be implied, and a sort of *systematic order* certainly is involved. Heron justifies this useful arrangement and undertaking by insisting that the combinations of the four elements air, fire, water and earth,[171] or of three of these, produce in some cases useful things that are indispensable for human life (ἀναγκαιοτάτας τῷ βίῳ τούτῳ χρείας),[172] in other cases marvels that cause astonishment. Note that at the end of the introduction he states that, 'these things [*scil.*, the issues concerning the void] having been *clarified*, we shall *next* describe' the marvels produced by the combinations of the elements (which combinations he had mentioned at the beginning and refers back to now): p. 28.28.11-4, τούτων δὴ διασεσαφισμένων ἐξῆς ... γράψομεν. Again, τάξις.

VI 4 *The* Automata

The proem to the *Automata* is quite interesting (ch. i, 338.3-342.10 Schmidt). Firstly, there is a brief reference quite similar to that in the *Pneumatica*, viz. to earlier writers (τῶν πρότερων) who have occupied themselves with αὐτοματοποιητική because of its wonderful

---

[169] Cf. above, text to n. 74, on Pappus who *Coll.* VIII concentrates on the theoretical aspect, while Heron (if we forget about his dissertation on the void) is only concerned with the production of miraculous effects.

[170] For this *topos* see above n. 117 *ad finem*, and text thereto. For the orderly presentation e.g. Schmidt at Nix and Schmidt (1900) 306 (with references): "[n]ach der Aufgabe folgt [each time] eine Art analytischer Betrachtung des Einzelnen und darauf die zusammenfassende Darstellung des Ganzen."

[171] Cf. the end of the dissertation of the void, 28.12-4 Schmidt, and see Gottschalk (1965) 116, also for the parallels in Philon mechanicus.

[172] Cf. above, n. 71.

effects. This is so because each part of mechanics is so to speak involved in αὐτοματοποιητική. So the mathematical *subdiscipline* it *belongs with*, viz. mechanics, is indicated *disertis verbis*. A descriptive overview of the *contents* of the treatise follows: there are moving automata and standing automata. Heron states that the former are described 'in the present book' (ἐν τούτῳ τῷ βιβλίῳ ... γράφομεν), the latter 'in the next' (ἐν ... τῷ ἑξῆς ... γράφομεν). So there is authorial authority for a *division into parts*, viz. into two books. In the Teubner edition the second book begins at ch. xx.

VI 5 *The* Catoptrica

The *Catoptrica* is extant in a presumably abridged version only, in a medieval Latin translation (by Willem van Moerbeke), and in the mss. is ascribed to Ptolemy and entitled *De speculis*. Quite a few isagogical issues are found here, some of which have helped to underpin the attribution to Heron:[173] *utility* (318.9 Nix and Schmidt, *dignum studio*; 318.18, *opportunitates necessarias*); the reference to predecessors (320.6-7, *puto necessarium esse accepta ab hiis qui ante nos descriptione dignificari*) which belongs with the *aim* of the work; the *orderly* arrangement, as is especially clear from the concluding chapters. This attribution to Heron, based on circumstantial evidence that is a bit thin, is of course far from certain and can be accepted only provisionally. On the other hand it is hard to come up with an alternative.

The introduction (316-24) is philosophical, or rather scientific, in an interesting way:[174] It first mentions the two senses through which wisdom is achieved according to Plato (reference, of course, to *Tim.* 46c-47e). A Platonizing and Pythagoreanizing description of the music of the spheres follows, and then something about the acoustic effects of the moving stars on the air. Next we have a *division into* three *parts* of the theory of vision: *opticum* (well presented by 'our' predecessors, esp. Aristotle), *dioptricum* (Heron[?] refers to another treatise of his in which this part has been treated at length),[175] and *katoptricum*. The last of these also needs treatment,

---

[173] See Schmidt at Nix and Schmidt (1900) 305-6. For Moerbeke's translation see Vanhamel (1989) 367-8.
[174] Compare Theon(?)'s introduction to the later version of Euclid's *Optica*, below, text to n. 193, and Ptolemy, below, text to n. 194.
[175] See below, Ch. VI 7.

not only because it can be useful, *utilis,* for purposes which provide fun (carnival mirrors), but also because it is *utilis* for *opportunitates necessarias* (see above), examples of which are provided. Treatment will be complete and, we may assume, *orderly* (*ut in nullo deficiat negotium*). The following chapters deal at some length with the properties under various circumstances of light and the visual ray.

VI 6 *The* Metrica

The proem of the *Metrica* (3-6.7 Schöne) starts with the 'traditional story' of the origin of geometry from measuring and dividing the land, a *useful* (χρειώδης) technique.[176] This utility led to a further development of the γένος, so that also solids were measured. This necessitated the finding of further theorems, many of which were discovered by Archimedes and Eudoxus (examples provided), though much remains to be done. Because of the *indispensability* of this discipline (ἀναγκαίας ... ὑπαρχούσης τῆς ... πραγματείας) Heron has decided to collect the *useful* things described by his predecessors (ὅσα τοῖς πρὸ ἡμῶν εὔχρηστα ἀναγέγραπται), and to add what he has discovered himself. He will begin with the measurings of planes (= book I). The proem to book II (p. 92-96.11) states that after the measurings of planes and surfaces of solids in the previous book (ἐν τῷ πρὸ τούτου βιβλίῳ), the measurings of various solids have to be dealt with: difficult and so to speak paradoxical inventions, ascribed to Archimedes by some historians (τινὲς ... κατὰ διαδοχὴν ἱστοροῦντες, 92.8-9).[177] However this may be, these inventions

---

[176] The *Geometrica,* though as we have noticed extant only as a Byzantine manual (above, n. 162) exhibits a few interesting introductory features. It has no less than three proems (4.172-76.13 Heiberg): the first without a heading, the second with the heading Ἄλλως (so this is an alternative to the first), and the third with the heading Ἥρωνος ἀρχὴ τῶν γεωμετρουμένων. To start with the latter: this is about the origin of geometry from the measuring of land, just as in the proem to book I of the *Metrica.* There is an extra bit, viz. that this useful practice started in Egypt and then spread to mankind as a whole; nevertheless the authenticity of the piece is in doubt, since it may be no more than a revised excerpt from the proem of the *Metrica.* I do not know that it is possible to put a date to the other proems, so shall ignore them here.

[177] To the best of my knowledge this is the only surviving reference to a *Successions* literature dealing with mathematics, though perhaps also another (but in my view less plausible) interpretation is possible, viz. 'historians [not necessarily of mathematics] dealing one after the other' with Archimedes. Synesius' remark about 'the great Ptolemy and the divine band of his successors', *Ad Paeonium de dono astrolabi* 5, at Terzaghi (1944) 2.139.1-2 (Πτολεμαίου

too have to be described, so that future users will find no lacunae in the present work. A few preliminary technicalities follow. The short proem (p. 140-42.2) of the third and last book, which deals with the division of planes and solids, states that the difference between the measuring and the division of places is not great. Parcelling out pieces of land in equal portions (or in unequal portions, when people deserve more) is *useful* and *indispensable* (εὔχρηστον καὶ ἀναγκαῖον). Nature herself has already divided up the earth in this way, and so have men. However, for division to be absolutely precise (and so equal, or just) one needs geometry, the only science which gives us proof that is indisputable.

A number of isagogical issues are again present: the *theme* or themes, also in relation to the work of predecessors and the *history* of the subdiscipline; the *division into parts*, i.e. books, for which there is authorial authority: the *systematic ordering* of these parts; the *relation* of metrics *to* the theoretical disciplines of mathematics, esp. geometry (and stereometry: note that Heron uses the first term only);[178] and *utility*, of course.

VI 7 *The* Dioptra

In the introduction to the *Dioptra*[179] (188-190.23 Schöne) we hear tones that by now must have become quite familiar. We hear of its manifold and indispensable uses (πολλὰς καὶ ἀναγκαίας ... χρείας), i.e. its *utility*, worked out in some detail in ch. 2: for daily life (πολλὰς ... τῷ βίῳ[180] χρείας), viz. its usefulness for irrigation, the building of walls etc.; for another mathematical subdiscipline, viz. astronomy (τὴν περὶ τὰ οὐράνια θεωρίαν) because it measures the distances between the stars, and deals with the sizes, distances, and eclipses of sun and moon; for geography; and for the arts of war. So we are informed of the *relation* of dioptrics to *other subdisciplines*. But, to return to ch. 1: Heron intends to treat what has been neglected by his predecessors, to formulate what has been said in a difficult way in an easier way (issue of *clarity*), and to correct mistakes that have been made. He will not do so in detail, as readers may look

τοῦ πάνυ καὶ τοῦ θεσπεσίου θιάσου τῶν διαδεξαμένων) is no more than a *façon de parler*, and perhaps taken too seriously by Neugebauer (1975) 2.873.
178 Cf. above, text to n. 112.
179 Written before the *Catoptrica*, cf. above, Ch. VI 5.
180 Cf. above, n. 71.

up what others have written and notice the differences themselves. A more important point is that others have used a variety of instruments with little result, while Heron will make use of a single instrument, the *dioptra*,[181] for the solution of many problems, and it will doubtless come in handy for other problems too. At the end of ch. 2 he tells us that *first* he will explain the construction of this instrument, and *next* set out its uses (χρείας again): an *orderly* and *systematic division into parts*.

VI 8 *A Theoretical Work: the So-Called* Definitiones, *i.e.* Τὰ πρὸ τῆς γεωμετρικῆς στοιχειώσεως

The next work to be discussed is the *Definitiones*, a Byzantine collection of abstracts, of which Nos. 1-132 are convincingly argued by Heiberg to derive from Heron. We do not know to what extent Heron's text was abridged. The Byzantine compiler added abstracts from his *Geometrica* (No. 133), from Euclid's *Elements* (No. 134), from (perhaps!) Geminus (No. 135), from Proclus *in Eucl.* I (Nos. 136-7, quite long), and from Anatolius (No. 138).[182] Here I shall of course restrict myself to the Heronian part of the collection. The short proem, dedicating the work to a certain Dionysius, has been preserved (p. 14.1-9 Heiberg). I find this section extremely interesting, not only because Heron formulates his *didactic purpose*, viz. to make the treatises of Euclid and others more easily comprehensible (εὐσυνόπτους, issue of *clarity*) to *students*, or because he says that his starting-point and whole *orderly arrangement* (τήν τε ἀρχὴν καὶ τὴν ὅλην σύνταξιν) will conform to the example set by 'Euclid the Elementarist', but especially in view of the general description of the work which is found at the beginning. This formula is τὰ ... πρὸ τῆς γεωμετρικῆς στοιχειώσεως τεχνολογούμενα, 'the systematic introduction *which comes before* the Elements of geometry'. It may well be the case that this so-called *Definitiones* and not the Commentary is the work on Euclid listed in the *Fihrist*,[183] but one cannot be sure.

---

[181] This instrument serves about the same purposes as the modern theodolite.

[182] The encyclopedia article of Mahoney (1972) contains nothing new compared with Heath (1921) 2.314-6. For the Anatolius paragraph in [Heron] see below, n. 228.

[183] Cf. below, pp. 125-6, complementary note 77 *ad finem*.

The formula τὰ πρὸ (the reading or study of ...) can be paralleled in both earlier and later authors, and is sort of giveaway formula indicating an introduction to an author or corpus, to a particular work, or to a discipline. Thrasyllus, about one generation before Heron, called his introduction to the collected works of Democritus Τὰ πρὸ τῆς ἀναγνώσεως τῶν Δημοκρίτου βιβλίων, 'What Comes Before the Reading of the Books of Democritus' (D. L. 9.41). Two centuries later Origen ends the lengthy introduction to his Commentary on John with the words, In Ev. Ioann. 1.88: 'here we shall end what comes before the reading in class of what has been written' (αὐτοῦ που καταπαύσομεν τὰ πρὸ τῆς συναναγνώσεως[184] τῶν γεγραμμένων). A slightly different formula, stating the position of the Pythagoran Golden Verses at the beginning of the philosophical curriculum, is found in Hierocles the Platonist's Commentary on this short poem: 'this is the aim and position of the Verses, to impress a philosophic character on the students before the other readings' (οὗτος μὲν ὁ σκοπὸς τῶν ἐπῶν καὶ ἡ τάξις, χαρακτῆρα φιλόσοφον πρὸ τῶν ἄλλων ἀναγνωσ-μάτων ἐνθεῖναι τοῖς ἀκροαταῖς, in Carm. aur. pr. 4 Köhler). The aim (σκοπός) is to turn the students into beginning philosophers, the order (τάξις) pertains to the fact that the Golden Verses are studied, in class of course, before all the other works that are eventually to be studied. Proclus is next; at in Remp. 1.1.5-7 Kroll (cf. ibid. 5.3-5) he gives the contents of a chapter as follows: 'On which and how many headings must be distinctly described before the reading in class of the Republic of Plato by those who wish to interpret it correctly' (περὶ τοῦ τίνα χρὴ καὶ πόσα πρὸ τῆς συναναγνώσεως τῆς Πολιτείας Πλάτωνος κεφάλαια διαρθρῶσαι τοὺς ὀρθῶς ἐξηγουμένους αὐτήν).[185] Finally, we may mention Ammonius Hermiae who at in De int. 1.24-6 Busse refers to his Prolegomena, or rather

---

[184] For συνανάγνωσις in Nicomachus see above, n. 69 and text thereto; also see below, n. 306 and text thereto.
[185] For details concerned with the practice involved see Mansfeld (1994) 245, index s.v. 'reading'. For Theon(?)'s parallel title see text to n. 195 below, and for the descriptive phrase in the proem of Aelius Theon's Progumnasmata see below, p. 122, complementary note 5. We may also recall the Hellenistic title of the work by Aristotle later called Categories, viz. Τὰ πρὸ τῶν τόπων α' (D. L. 5.24; same title in the Theophrastus' catalogue at D. L. 5.50), see Frede (1983) 12-8 = (1987a) 17-21: the work was considered to be preliminary to the Topics; see also cf. De Libera and Segonds (1998) xv n. 23. A similar idea is behind the characterization, in the famous scholium at the end of the treatise in a number of mss., of Theophrastus' so-called Metaphysics as προδιαπορίαι τινὲς ὀλίγαι of the entire discipline, viz. metaphysics; see Laks and Most (1993) xvi-xviii.

Prolambanomena, in the following words: 'in the preliminaries to the reading in class of the Categories', ἐν τοῖς προλαμβανομένοις τῆς συναναγνώσεως τῶν Κατηγοριῶν.[186]

The terminus technicus προτεχνολογούμενα, and forms of the verb τεχνολογεῖν plus πρό are rare and mostly found in late authors.[187]

I believe that the formula τὰ ... πρὸ τῆς γεωμετρικῆς στοιχειώσεως τεχνολογούμενα in the proem of the *Definitiones* is the *original* Heronian *title*, a belief that is underpinned by no less than two self-references in the *Definitiones* to a similar (though now lost) *Introduction to Arithmetic* by Heron, viz. Τὰ πρὸ τῆς ἀριθμητικῆς στοιχειώσεως, 'What Comes Before the Elements of Arithmetic' (p. 76.23 and 84.18).

We note that utility is *not* mentioned; in fact the work is wholly theoretical, not practical, as Heron's other works are. On the other hand, that the work in facts is meant to be *useful* as a general introduction to geometry is beyond doubt.

---

186 Cf. Olymp. *Prol.* 1.8, 1.26, 2.9-10, 14.11-2, 25.22-3 Busse.

187 Eus. *in Psalmos*, Migne *PG* 23, 1001.35 (ἐν τοῖς προτεχνολογουμένοις) and 1072.22-3 (ἐν τοῖς προτεχνολογουμένοις τῶν ψαλμῶν), Ammon. *in Isag.* 21.7 Busse (προλεγόμενα ἤτοι προτεχνολογούμενα), Stephanus *Ethn.* 47.20-1 Meineke (ἐν τοῖς τῶν ἐθνικῶν προτεχνολογήμασιν εἴρηται), beginning of excerpt from the προθεωρία of Severus' *Epithalamium* at Phot. *Bib.* cod. 243, 366b Bekker (ἴσως μὲν ἂν τῷ περίεργον εἶναι δόξειε τὸ πρὸ τῶν ἐπιθαλαμίων τεχνολογεῖν); see further Mansfeld (1994) 10 n. 2.

CHAPTER SEVEN

THEON(?)'S PREFACE TO EUCLID'S *OPTICA*

As we have seen above Theon of Alexandria published a revised version of Euclid's *Elements*. We also have a revised version of the *Optica*[188] which has been traced to Theon, though unlike the edition of the *Elements* it is not designated in this way in the mss. This revision is prefaced by an introductory essay, 144.1-55.2 Heiberg.[189] Heiberg argued that this is the authorized report by a pupil of his teacher's introduction to his exposition ("Lehrvortrag") of the work.[190]

This piece is interesting in various ways. The first of these is that isagogical questions are not at all at issue explicitly, though we may infer that the *authenticity* of the ἐπιγραφή was regarded as unproblematic. Moreover the report may well be incomplete, the pupil (or a later *scriba*) preserving only what he believed to be really interesting.

The second point of interest is that the lecturer very firmly places Euclid's treatise in the context of physics and sense-perception.[191] The original version of Euclid's *Optica* is the most

---

[188] Both versions ed. Heiberg (1895). Heiberg (1882) 139 bases the ascription to Theon on a scholion in *Paris. gr.* 2468: τὸ προοίμιον ἐκ τῆς τοῦ Θέωνός ἐστιν ἐξηγήσεως. Because this ms. was written in 1565, the ascription has little or no authority; we may observe that the scholion is not (!) found in Heiberg's edition of the scholia to the later version at Heiberg (1895) 251 ff. Even so, Heiberg's view was accepted by authorities such as Heath (1921) 1.441, Ziegler (1934) 2079, Neugebauer (1975) 2.893, and Knorr (1989) 452 n. 17; also by Fraser (1972) 1.389. Toomer (1976b) 322 writes that "there is *no direct evidence* [my italics] ... that Theon was responsible for this version, though he remains the most likely candidate".

[189] Preliminary ed. with facing German transl. Heiberg (1882) 138-45.

[190] Heiberg (1882) 138-9, 145-6: the words ἀποδεικνύς, ἐκόμιζε (144.1), ἔφασκεν (144.9) do not apply to Euclid but to the lecturer: an example of what came to be called ἀπὸ φωνῆς, for which practice see Richard (1950). For earlier evidence concerning the noting down of a master's lectures see Sedley (1989) 103-4, and Dorandi (1997b) 46, 48, who argues that certain works by Philodemus are ἀπὸ φωνῆς [*scil.*, of Zeno of Sidon]; for similar evidence concerning the Sceptical Academy see Mansfeld (1994) 193. For Marinus see below, Ch. VIII.

[191] For the physicalist aspects of the introduction to Heron(?)'s *Catoptrica* see above, Ch. VI 5. Even purely geometric optics fails to avoid physics

purely mathematical of all extant ancient treatises on, or accounts
of, optics and vision, though his visual rays are real physical
entities. Greek optics and theories of vision are in several ways
defective; naturally, light is not given the predominant role it plays
since the discoveries of ibn al-Haytham/Alhazen, Kepler, and
Descartes, but as a rule is only a necessary partner of the (e.g.,
fiery, or pneumatic) rectilinear visual rays, or of the visual cone
which, depending on the particular theory at issue, may be
formed by the rays themselves or by the medium that is influenced by the agent of seeing. These rays or this cone, issuing from
their base in or upon the eye, are so to speak a kind of fingers, or
sticks, which touch the objects that are seen and then report
back.[192] In conformity with the mainstream tradition of ancient
geometrical optics Theon(?) too posits that the eye sends out a cone
of straight visual rays.[193] In this context, however, it is important to
note that the great Ptolemy in his *Optica*—only books II-V are extant
in a medieval Latin translation from the Arabic, while the end of
book V is lost too—refined this traditional geometric optics even
further, but also revised it and far more straightworfardly placed it
in a physical setting. On the one hand he argued that the rays in
the cone form a continuum, and so turned them into mere abstrac-
tions. On the other he payed proper attention to the indispensable
role played by the illumination of the sensible object and the
qualities such an object must have in order to reflect illumination,
to the perception of the proper object of vision, colour, and via
colour to the apperception of other qualities of the object. And he
performed experiments to underpin his theoretical views.[194]

Several arguments in support of Euclid's doctrine of visual
perception are offered by Theon(?) in the course of his exposition,
e.g. that the eye is globular, not hollow like the ears, nostrils, and
mouth, as it would be had it been a purely receptive organ. We

---

altogether, see Lindberg (1976) 11-7 on the mathematicians, and on Greek
optics in general the impressive overview of Crombie (1994) 1.155-76, who
demonstrates that the theories gradually came to include more and more
physics and physiology.

[192]   See below, pp. 127-8, complementary note 192.

[193]   As is postulated in the first definition of Euclid's *Optica* in both
recensions. Also other matters explained in Theon(?)'s introduction pertain
to the definitions.

[194]   Ed.: Lejeune (1956) 11; see further Lejeune (1947), Lejeune (1948) 38-41,
65-6, on the lost book I of the treatise, Neugebauer (1975) 2.894-6, Simon (1988)
83-91, and esp. Smith (1988), Crombie (1994) 1.162-70.

may thus infer that he wanted to provide a stronger, or at least more elaborate, physicalist context for Euclid's treatise in order to make it look less old-fashioned.

Most important from our point of view, thirdly, is the fact that the title of the piece in the mss. is *Τὰ πρὸ τῶν Εὐκλείδου Ὀπτικῶν*, 'What Comes Before the Optics of Euclid'. There is no independent proof either *pro* or *contra* the assumption that this title is original, but what should be noticed in favour of its being authentic is that the designation 'What comes before ...' (Τὰ πρὸ ...) in this context can be paralleled quite early, as we have seen above.[195] So whoever gave the introduction to the so-called *recensio Theonis* of Euclid's *Optica* its present designation was well-informed, and placed the piece in the sub-genre to which it belongs.

---

[195] Ch. VI 8.

CHAPTER EIGHT

MARINUS ON EUCLID'S *DATA*

Proclus' pupil Marinus of Flavia Neapolis (Nablous) is not only the author of the well-known *Encomium* written after his teacher's death, but also of a short preliminary piece dealing with the *Data* which is less familiar to students of Neoplatonism.[196] *Pace* Menge (and the misleading title of Michaux's little monograph) what we have here is not a 'commentarius'.[197] Though the first hand in *Vaticanus graecus* 204 (9th-10th cent.) has ὑπόμνημα εἰς τὰ δεδόμενα εὐκλείδους ἀπὸ φωνῆς μαρίνου φιλοσόφου, the rubricated correction προθεωρία κτλ. for ὑπόμνημα κτλ. by a much later hand is certainly apposite. Perhaps the commentary in the proper sense of the word, viz. the part pertaining to the work itself,[198] has been lost,[199] the προθεωρία (or προλεγομένα, as a later ms. has it) being the only part that has been preserved. Alternatively, Marinus used the Commentary of Pappus to which he refers *ad finem*, and did not bother to have his comments on the work itself (and his remarks on the Commentary) taken down by one or more of his pupils.[200] We may further observe that the piece that is extant conforms to the section 'before the work', viz. the first part, of the division *ante opus* (i.e. the prolegomena) and *in ipso opere*, 'on the work itself' (i.e. the commentary proper) of a commentary, a division said to be

---

[196] Ed. Menge (1896b); see further the encyclopedia article of Schissel von Fleschenberg (1930) and the monograph of Rome's pupil Michaux (1947). Several works by Marinus have been lost. He was Damascius' teacher in geometry, arithmetic, and the other mathematical disciplines, see Dam. *Isid. ap.* Phot. *Bibl.* cod. 181, 126b-27a Bekker (p. 199 Zintzen), γεωμετρίας δὲ καὶ ἀριθμητικῆς καὶ τῶν ἄλλων μαθημάτων Μαρῖνον ... ἔσχε διδάσκαλον. According to Elias *in Isag.* 28.9 Busse he said 'I wish everything were mathematics', διὸ καὶ ὁ φιλόσοφος Μαρῖνος ἔφη· εἴθε πάντα μαθήματα ἦν. For his interest in astronomy see below, n. 222 and below, p. 129, complementary note 260.

[197] Cantor (1907) 282, followed by Schissel von Fleschenberg (1930) 1761, rightly speaks of a "Vorrede". Michaux (1947) 67 ff. agrees.

[198] See below, n. 201 and text thereto.

[199] Thus Schissel von Fleschenberg (1930) 1761, Michaux (1947) 71, Sambursky (1985) 17.

[200] According to Dam. *Isid. ap.* Phot. *Bibl.* cod. 242.146 (p. 198 Zintzen) he 'copied the views of the commentators and reserved a copious amount of notes for his own use', ὑπομνήματα καταλείπων ἑαυτῷ καὶ ἀποθησαυριζόμενος.

standard by Aelius Donatus (mid-4th cent. CE) in his Commentary on Virgil.[201]

Marinus right at the start lists three (or rather four) preliminary questions in the appropriate scholastic way:[202] the explanation of the *title* which involves that of the *theme*, since the term δεδομένα, which has to be defined, functions both as title and theme; next the *utility* of the discipline which studies this subject; and thirdly *under what scientific discipline* it has to be subsumed.[203] The discussion of the theme, quite appositely, starts with a *historical* overview, with *inter alia* references to Apollonius' *Inclinationes*[204] and his 'general work', i.e. probably the treatise called *De principiis mathematicis* by Heiberg,[205] to Ptolemy, and to Diodorus (234.15-36.1 Menge),[206] and branches out into a lengthy enquiry into the proper definition of the term δεδομένον (see below). The χρήσιμον is discussed 252.20-54.4: knowledge of the *Data* is indispensible for Analysis.[207] The importance of Analysis, the author continues, for the disciplines of (pure) mathematics and related disciplines such as optics and canonics 'has been defined elsewhere' (ἐν ἄλλοις διώρισται).[208] In this other work Marinus, as he says, has also pointed out that Analysis is the discovery of proof, i.e. a heuristic method, and how much it contributes to the finding of similar proofs, and that it is much more important to be capable of using Analysis than to be already in possession of numerous individual proofs. Pappus had restricted the utility of Analysis to the solution of problems set to

---

[201] For this distinction and its applications see Mansfeld (1994) 43, 44, 49, 116, and cf. above, text to n. 198, below, text to n. 275.

[202] On isagogical questions in Marinus see Schissel von Fleschenberg (1930) 1761-2, who speaks of the "Bestand der Bucheinleitung" as part of this introduction (cf. below, n. 250); note however that he is unaware of the nature and existence of the isagogical scheme itself. He is followed by Michaux (1947).

[203] 234.1-3 Menge, Πρῶτον δεῖ θέσθαι τί τὸ δεδόμενον· ἔπειτα τί τὸ χρήσιμον τῆς περὶ τούτου πραγματείας, εἰπεῖν· καὶ τρίτον ὑπὸ τίνα ἐπιστήμην ἀνάγεται.

[204] Belonging to the domain of Analysis, see Pappus *Coll.* VII, 2.636.22; above, Ch. II 2.

[205] Apollonius Fr. 51 Heiberg; see Heath (1921) 2.192-3.

[206] Possibly the Diodorus mentioned by Pappus *Coll.* 1.246.1; see Heath (1921) 1.358, 2.287, 2.359. Reference to 'Archimedes' predecessors' at 244.1-2, to Archimedes himself at 248.3.

[207] πρὸς .. τὸν ἀναλυόμενον λεγόμενον τόπον (cf. above, nn. 25 and 27). This agrees with the view underlying Pappus' sequence in *Coll.* VII, viz. that the *Data* are the first analytic work to be studied.

[208] One would very much like to know more.

students;[209] Marinus argues that the solution of problems is the main thing.

As to the issue *to what section of a discipline* the work belongs Marinus states (254.5-16) that because of its utility for all disciplines of the above kind it does not belong with a single particular subdiscipline, but with mathematics as a whole (εἰκότως ἂν ῥηθείη ἀνάγεσθαι οὐχ ὑπὸ μίαν τινὰ ἐπιστήμην, ἀλλ᾽ εἰς τὴν καθόλου λεγομένην μαθηματικήν). General mathematics is then defined. Euclid wrote the *Data* with this most useful cognitive purpose in mind, so he is rightly called 'Elementarist'. For before mathematics as a whole, so to speak, he has placed elements and introductions: of geometry in the thirteen books (*scil.*, of the *Elements*), of astronomy in the *Phaenomena*, also of optics and canonics. More especially, in the book in front of us now he has provided the foundation for Analysis (στοιχείωσιν ἀναλυτικήν). Further praise of Euclid follows. This section as a whole (254.5-27) somehow mirrors the well-worn scheme formulated by Quintilian *Inst.* 2.15.5 as *de arte, de opificio, de opere*, which also forms the backbone of Proclus' introduction to Euclid's *Elements*.[210] We may of course safely assume that Marinus was familiar with Proclus' Commentary on *Elements* book I, with its twofold introduction.

At the end (256.10-22) Marinus discusses, or mentions, further issues. First, as a fourth (or rather fifth) preliminary question the *division* of the treatise *into parts*.[211] Two different divisions are given, the first of which distinguishes four parts according to the species of δεδομένα: the πρῶτον ... τμῆμα [note that τμῆμα by now is an isagogical terminus technicus] deals with the δεδομένα κατὰ λόγον,[212] the δεύτερον with those τῇ θέσει,[213] and the next with those τῷ εἴδει.[214] The fourth species, that of the μεγέθει δεδομένα,[215] though simple (ἀπλοῦν), is parcelled out among the others (κατέσπαρται ... μερικῶς), mostly in the third section.

---

[209] See above, text to n. 26; below, n. 219 and text thereto, and below, p. 123, complementary note 26. Also see Knorr (1986) ch. 8.

[210] See Van Berchem (1952) 81, Mansfeld (1994) 39 with n. 60 (where further references to the literature). For another example see below, text to n. 275.

[211] Michaux (1947) 17, 47 incorrectly views this as an appendix instead of an integral part of the scheme.

[212] Cf. *Data*, def. 2.

[213] Cf. *Data*, def. 4, 8.

[214] Cf. *Data*, def. 3.

[215] Cf. *Data*, def. 1, 6, 7, 8, 9-12.

A *systematic sequence* is involved here (and so, of course, an *order of study*—a further isagogical issue, viz. the fifth or rather sixth): Euclid, Marinus says, began with the λόγῳ and θέσει δεδομένα, since the δεδομένα τῷ εἴδει are composed of these.

A second, alternative (καὶ ἄλλως) *division* into four *parts* is also described, viz. according to magnitudes in general, lines, planes, and theorems concerning circles. A similar τάξις (*systematic sequence*) was applied by the author (i.e. Euclid) also to the definitions, or hypotheses, of the book. Interestingly enough, this division is *grosso modo* the same as that of Pappus' summary of the *Data* in the *Collectio*, though Marinus worked with a text which differed to some extent from that used by Pappus.[216]

Finally, a sixth or (rather seventh) issue is brought into play, viz. Euclid's '*method of instruction*' (τρόπος τῆς διδασκαλίας).[217] This according to Marinus is *not* κατὰ σύνθεσιν but κατὰ ἀνάλυσιν, 'as Pappus convincingly demonstrated in his [for us lost] Commentary (τοῖς ... ὑπομνήμασιν) on the book'.[218] This remark is somewhat surprising, since Pappus at *Coll.* 2.624.8-11 Hultsch affirms that the method of *Euclid*, Apollonius and Aristaeus is about 'Analysis *and synthesis*', κατὰ ἀνάλυσιν καὶ σύνθεσιν. Nevertheless it seems to be beyond doubt that it is a view of Pappus which forms the background of Marinus' stance, though in a way which looks a bit idiosyncratic.[219] Even so, this reference is not only important because it constitutes our only evidence for Pappus' Commentary on the *Data*, but also because we may believe, or so I think, that part of Marinus' discussion concerning the first isagogical issue, that of the various meanings of δεδομένον, to some extent at least depends on Pappus, one of 'the commentators he excerpted'.[220] The historical information included there may well go back to him too;

---

[216] Michaux (1947) 48-51.

[217] See below, p. 128, complementary note 217.

[218] In the *Collectio* Pappus includes the *Data* in the domain of Analysis, see above, Ch. II 2; note that in this work the *Data* are merely summarized, not discussed or commented upon. Heiberg (1882) 173 already pointed out that Marinus' remark cannot pertain to *Coll.* 2.638-40.1 Hultsch. For the speculative solution of Jones see above, n. 33.

[219] Also cf. above, text to n. 209. Perhaps Marinus exaggerated a point of view expressed by Pappus in the lost Commentary resembling that quoted above, text to n. 26. Knorr (1986) 357-60, who appositely cites Arist. *EN* 3.3.1112b15-27, argues that Pappus' description is indebted to philosophical views concerning analysis and synthesis. Also cf. below, p. 123, complementary note 26.

[220] See above, n. 200.

one only has to think of the introductory paragraph of Pappus' extant Commentary on *Elements* book X, with its references to the Pythagoreans, Theaetetus, Eudemus and Apollonius. The careful distinction of the various views pertaining to the meaning and proper definition of the term δεδομένον and its relation to other concepts (viz. τεταγμένον, γνώριμον, ῥητόν, πόριμον, ἄτακτον, ἄγνωστον, ἄπορον, ἄλογον) which takes up most of Marinus' tract resembles Pappus' careful conceptual discussion of the 'rational' and the 'irrational' and of other technical terms in the first part of the Commentary; but I cannot go into this matter here.

No discussion of the term δεδομένον is found in the *Collectio*, but one may observe that the synonymous term δοθέν is briefly explained in the desciption of Analysis at 2.636.7-12. It is a mathematical *terminus technicus* (ὃ καλοῦσιν οἱ ἀπὸ τῶν μαθημάτων δοθέν): 'In the case of the problematic kind, we assume the proposition as something we know, then, proceeding through its consequences, as if true, to something established, if the established thing is possible and obtainable [ποριστόν, cf. Marinus' πόριμόν], which is what mathematicians call "given", the required thing will also be possible.'[221] Marinus goes his own way, but what he tells us is nevertheless indebted to at least one of his predecessors.[222] We have seen above, moreover, that the second division into parts of the contents of the *Data* mentioned by him is entirely similar to the overview given by Pappus in the *Collectio*, and it is only to be expected that an overview, or division, of this nature was also to be found in Pappus' lost Commentary.

---

[221] Transl. Jones (1986a) 1.84. It will be clear that this passage cannot have been Marinus' source.

[222] Another reference to Pappus by Marinus exists, viz. in the for the most part unpublished scholia on Theon's *Little Commentary* (cf. below, n. 261 *ad finem*) which are the remains of a late, possibly Alexandrian Commentary according to Tihon (1976). Here we read that 'the philosopher Marinus says that Pappus spoke about the parallaxes in conformity with what is been proved in book V of the *Suntaxis*', ἀκολούθως τοῖς ἐν τῷ πέμπτῳ τῆς Συντάξεως δειχθείσι τὸν Πάππον φησὶν ὁ φιλόσοφος Μαρῖνος τὰ περὶ τῶν παραλλαξέων λέγειν κτλ. The text is published by Tihon *ibid*. 183; for its interpretation see *ibid*. 173-5. For Marinus' interest in Ptolemy also see below, p. 129, complementary note 260.

CHAPTER NINE

PTOLEMY'S PREFACES

IX 1 *The* Mathèmatikè Suntaxis

Ptolemy, about a generation earlier than Galen, as we shall see planned and executed his works very carefully.[223]

The headings of the first two chapters of book I of the *Mathèmatikè Suntaxis*[224] are προοίμιον and περὶ τῆς τάξεως τῶν θεωρημάτων, both in the *pinax* and in the work itself.[225] These two chapters taken together may be viewed as forming the introduction to the whole treatise.[226] In the first chapter, which dedicates the work to his standard dedicatee Syrus, Ptolemy advises us about the place and value of mathematics. He first accepts the division of the sciences into the theoretical (which provides πλειστὴν ὠφέλειαν, I.1.4.15

---

[223] I omit most of the *Optica* of which the first book is lost (above, text to n. 194), the *Inscriptio Canobis* which is without introduction, and the *Planisphaerium*, which though dedicated to 'Jesurus' (II.227.1 Heiberg) i.e. Σύρος (originally ὦ Σύρε, or υἱὲ Σύρε?) lacks a proper introduction. The other minor astronomical works will be adduced whenever profitable; in themselves they do not add much to what can be learned for our purposes from the *Suntaxis* or *Apotelesmatica*. Ed. Heiberg (1907): *Phaseis* 1-67, *Hypotheseis* 70-145 (book II in German, from the Arabic), *Inscriptio Canobi* 148-55, *Procheiroi canones* 159-85 (the introduction alone, i.e. not the tables [cf. below, n. 261], much altered in later times), *Analemma* 189-223 (Greek fragments and medieval Latin transl. from the Arabic), *Planisphaerium* 227-59 (medieval Latin transl. from the Arabic), *Fragmenta* 263-70. On Ptolemy see e.g. Ziegler *& al.* (1959), Lloyd (1973) 113-35, Toomer (1975).

[224] Note the self-references at *Hyp.* 2.70.1-2 Heiberg (ἐν ... τοῖς τῆς Μαθηματικῆς συντάξεως ὑπομνήμασιν) and *Geogr.* 2.195.25-6 Nobbe (ἀπεδείξαμεν ἐν τῇ Μαθηματικῇ συντάξει).

[225] See below, pp. 128-9, complementary note 225.

[226] The Platonizing ingredients of ch. 1 of book I have been discussed by Taub (1993) 19-37; Boll (1894) 66 ff., who emphasized the Peripatic background but also pointed at Platonic and Stoic ingredients in Ptolemy, remains useful. Ptolemy's ranking of mathematical astronomy looks like an emendation of Aristotle's view that it is the mathematical discipline which comes closest to philosophy, *Met.* Λ 8.1073b3-8. Hadot (1984) 256 writes: "(à) cause de ce mélange d'éléments stoïciens, péripatéticiens et platoniciens [viz. as analyzed by Boll] dans la philosophie de Ptolémée, je n'exclurais pas la possibilité qu'il ait été un moyen-platonicien". This goes a shade too far: an interest in philosophy or the use of philosophical ideas do not make a person a philosopher (cf. below, n. 325, text to n. 355).

Heiberg) and the practical advocated by what he calls the 'genuine philosophers',[227] and next to what—not improperly—he calls 'Aristotle's division of the theoretical science into physics, mathematics, and theology' (cf. *Met.* E 1.1026a18-9, K 8.1064b1-3).[228] Next he argues that mathematics is not only the most scientific and secure of the theoretical disciplines, but also makes a major contribution (συνεργεῖν, I.1.7.4) to the other two, and especially to theology insofar as it puts the study of the heavens and the cosmic order on unshakeable foundations. Without mathematics theology is guesswork, its object of study (*scil.*, the divine itself) being 'entirely invisible and out of reach', and so is physics because of the 'unstable and opaque nature of matter'; it is therefore not to be expected that the philosophers will ever agree among themselves, that is to say about issues in theology and physics.[229] The

---

[227] Boll (1894) 70 n. 3 aptly cites the bipartite division at Arist. *Met.* α 1.993b19-21 (authenticity not in doubt), and ps.Plut. *Plac.* prooem. 874F (~ Aët. *DG* 273.25-74.5 Diels, not entirely happily positioned as Thphr. Fr. 479 FHSG), bipartite division according to 'Aristotle, Theophrastus and the majority of the Peripatetics'. Add D. L. 5.28, and for Aristotle himself (?) *Protr.* Fr. B 32 Düring at Iambl. *Protr.* 37.26-38.3 Pistelli. The parallels in ps.Plutarch and Diogenes show that by Ptolemy's time this had come to be seen as a standard Aristotelian view. This identification was already proposed by Theon *in Synt.* 320.6-8 Rome: λέγει δὲ τοὺς ἐκ τοῦ Περιπάτου, ἐπεὶ καὶ μετ' ὀλίγα τοῦ Ἀριστοτέλου μνημονεύων κτλ. Formulas resembling Ptolemy's expression οἱ γνησίως φιλοσοφήσαντες are quite common and occur in authors of various colours, both early and, mostly, late; they are first found in Plato *Phd.* 66b (τοῖς γνησίως φιλοσόφοις), *Resp.* 473cd (ἐὰν μή ... φιλοσοφήσωσι γνησίως τε καὶ ἱκανῶς, passage quoted Stob. *Flor.* 4.1.107). Also cf. e.g. Philo *Prob.* 3, ὅσοι δὲ φιλοσοφίαν γνησίως ἠσπάσαντο, the Pyrrhonist Sextus *M.* 1.280, and the Stoic Epictetus *Diss.* 3.26.23, οἱ γνησίως φιλοσοφοῦντες, Iambl. *Protr.* 63.30 Pistelli, τοῖς γνησίοις φιλοσόφοις. Somewhat different Procl. *Hypot. Astr.* ch. 1.1.2 Manitius, τόν γε ὡς ἀληθῶς φιλόσοφον, clearly echoing Plato's formula (*Phd.* 83b6, *Resp.* 376b1, 485e1, 490d6, 540d4), also at *in Remp.* 1.57.22 Kroll.

[228] Parallels for this division of philosophy including the tripartite subdivision of its theoretical part are to be found e.g. in Alcin. *Did.* chs. 3 and 7 (153.43-54.5 + 160-42-61.1 Hermann) as a Platonic doctrine, in an excerpt from Anatolius (ἐκ τῶν Ἀνατολίου) *ap.* [Heron] *Def.* § 138.1, 4.160.9-12 Heiberg (explicit attribution to Aristotle here; note the final words: μάλα σαφῶς καὶ ἐντέχνως φιλοσοφίαν οὖσαν τὴν μαθηματικὴν ἀποδείκνυσιν), and as Platonic doctrine again in the late Neoplatonists: Ammon. *in Isag.* 11.22-4 Busse, *in Cat.* 5.4-5 Busse, David *Prol.* 5.6-8 Busse (cf. *ibid.* 65.11-2,) David (Elias?—but see Ouzounian [1994]) *in Cat.* 115.18-9 Busse). Cf. also ps.Gal. *Part. phil.* §§ 1.1 + 3.1, 4.1 (explicit attribution to Aristotle, Plato's view of mathematics being different) and Joan. Damasc. *Dial.* rec. fusior § 3.28-31, § 66.16-9, rec. brev. § 49.17-9. A parallel for the subdivision of theoretical philosophy is in Anatolius' pupil Iamblichus, *CMSc.* ch. 28.

[229] I.1.6.16-7, ὡς διὰ τοῦτο μηδέποτε ἂν ἐλπίσαι περὶ αὐτῶν ὁμονοῆσαι τοὺς φιλοσοφοῦντας. Clearly Ptolemy is well informed about the διαφωνία of the

68 CHAPTER NINE

mathematical study of the divine phenomena also contributes in a most important degree to ethics, by rendering the souls of its practitioners similar to the equality, well-orderdness, symmetry, and modesty of the divine—a clearly Platonic touch.[230] This is the science Ptolemy will pursue systematically and to the best of his ability, briefly recording the findings of predecessors and unavoidably adding what has to be added.[231] So here the *intention of the author* is described in a way that is unmistakable. Furthermore, it is understood that only those students who have already made some progress in mathematical astronomy will be able to follow what is to be found in the treatise (οἱ ἤδη καὶ ἐπὶ ποσὸν προκεκοφότες δύναιντο παρακολουθεῖν, I.1.8.8-9). Anyhow 'everything useful for the study of the heavens will be set out in proper order' (ἅπαντα τὰ χρήσιμα πρὸς τὴν τῶν οὐρανίων θεωρίαν κατὰ τὴν οἰκείαν τάξιν, I.1.8.11-2).

It is clear that several isagogical issues are used here in an elegant and unpedantic way: the πρόθεσις of the author, as we have noticed already, the *position* of mathematics and mathematical astronomy vis-a-vis other theoretical sciences and practical

philosophers as demonstrated for instance in the *Placita* literature and the works *On Sects*. For his physicalist approach to astrology see the next section. In a comparable vein Nicomachus argues that the study of number is an indispensable contribution to physics, *Ar.* 1.23. 6 ff. at 65.17ff. Hoche. The sceptic view that physics is impossible because matter is in flux, and theology because the divine cannot be known seems to be traditional. It is formulated in a somewhat different way at David *Prol.* 5.13-7 Busse: τὰ ὄντα ἐν ῥοῇ καὶ ἀπορροῇ εἰσι καὶ στάσεως οὐδεμιᾶς τυγχάνουσι (Platonism without Forms, cf. Arist. *Met.* A 6.987a32-b1, Γ 5.1010a8-15, esp. M 4.1078b12-7, and see below, p. 127, complementary note 119 *ad finem*), and τὰ θεῖα αἰσθήσει οὐ καθυποβάλλονται, τὰ δὲ αἰσθήσει μὴ καθυποβαλλόμενα γνώσει οὐχ ὑποπίπτουσι, τὰ θεῖα ἄρα ἄγνωστά εἰσι (echoing Protagoras' famous dictum on the gods, 80B4 DK, cited e.g. by the Neopyrrhonist Sextus, *M* 9.55-6, and by Diogenes Laërtius 9.51, who treats Protagoras as a proto-Sceptic). These arguments (for which also see David *Prol.* 59.26-32 Busse, esp. τὰ θεῖα ἄτε δὴ ἀόρατα ὄντα καὶ ἀκατάληπτα εἰκασμῷ [cf. Ptolemy] μᾶλλον γινώσκονται ἤπερ ἀκριβεῖ γνώσει) are answered in a way different from Ptolemy's *ibid.* 5.31-6.21. Explaining the maxim ἀγεωμέτρητος μηδεὶς εἰσίτω attributed to Plato from the 4th cent. CE (see Swift Riginos [1976] 138-40) David also writes that 'mathematics contributes to the knowledge of theology', συμβάλλεται δὲ εἰς εἴδησιν τῆς θεολογίας τὸ μαθηματικόν, οὗτινος μέρος ἐστὶν ἡ γεωμετρία, *ibid.* 57.21-2; explained *ibid.* 59.12-23, with references to [Plato] *Epin.* 992a and Plot. *Enn.* 1.3.3. That mathematics (also in the sense of mathematical astronomy) contributes to physics and theology is of course Plato's doctrine in the *Timaeus*, and Aristotle's e.g. in the *De caelo* and *Met.* Λ.

[230] In a similar way Nicomachus grows eloquent about the side-effects on morality of the study of numeric ratios, *Ar.* 1.23.5 at 65.13-6 Hoche.

[231] Cf. above, n. 117.

science, the *utility* of theoretical science, mathematics, and especially mathematical astronomy,[232] the latter being useful not only
for the study of theology and physics but also for higher ethical
purposes, the *aim* of the present study, viz. to teach mathematical
astronomy in the best possible way[233] (the *historical* contributions of
others moreover will not go neglected), and the *order of study* as
well as the *qualities required of the student*, for students must to some
degree be prepared.[234] Perhaps Ptolemy also had Arist. *E N*
1.3.1095a11-3 in mind.[235]

The next chapter deals with 'the *order* of the theorems';[236] we
may observe that here too an isagogical question is involved.
Ptolemy however in this passage does not describe the contents
book by book.[237] Rather, he gives a *division into* two, three, or six
parts of the work as a whole, depending on how one counts (for
convenience I have added book and chapter numbers). Note that
the whole arrangement of these parts and sub-parts is perfectly
systematic and orderly, and that again and again Ptolemy
reminds his readers of this fact. Most of the time moreover the
contents of a previous book are summarized at the beginning of the
next.[238] Yet the division into books is so to speak overruled by
divisions of another kind.

The first of the parts into which the work as a whole after the
introductory section is divided, corresponds (*1*) to book 1.3-8, since
the general (καθόλου) relation of the earth to the heavens comes

---

[232] Utility also emphasized in the epilogue, I.2.608.7.

[233] Cf. *Hyp.* 70.11 ff., where the same claim is made for a simpler
treatment.

[234] See Toomer (1984) 6, who points out that this means a knowledge of
elementary geometry, 'logistic' i.e. calculation as taught at an elementary
level, and spherics (Euclid, Autolycus, Theodosius).

[235] Quoted n. 10 above.

[236] Cf. the enumeration in the proem of the *Can.* of the tables to be discussed, 159.14 ff. (οἱ ... πρῶτοι, οἱ ... ἐφεξῆς , etc.)

[237] For this see Toomer (1984) 5-6, who states that "the order of treatment
of topics ... is completely logical ". Note anyway that the division into books is
original (see below, nn. 238 and 241), cf. e.g. the first sentence of book II,
I.1.87.14, διεξελθόντες ἐν τῷ πρώτῳ τῆς Συντάξεως κτλ.

[238] So also at *Phas.* book II, with explicit reference to the lost first book
(3.15-6, ἐν τῇ κατ᾽ ἴδια συντάξει τῆσδε τῆς πραγματείας), at *Hyp.* book II (111.2 ff.),
and at *Opt.* book II. But this is not the case in the *Harmonica*, though this
treatise too is very systematic; see Düring (1930) xcvi-vii. The full-fledged practice itself is first found in the historians, e.g. Polybius and Diodorus Siculus
(on whom see below, p. 122, complementary note 11), see Birt (1882) 464-81,
Mutschmann (1911) 94-6, Van Sickle (1980) 7-8, and on Polybius Lorenz
(1931).

first in this treatise, προηγεῖται. The particular (κατὰ μέρος) *topics* are
next:[239] the first (πρῶτον, *2a*) of these fills a section on the ecliptic
corresponding to book 1.12-16, and is followed by one (*2b*) on the
regions of the world we inhabit corresponding to book II. Treatment of these issues will make the study of what is to follow easier
(again the *order of study*, this time for the contents of the treatise
itself).[240] Secondly (δεύτερον, viz. of the individual topics) a section
(*3*) on the sun and moon corresponding to books III-VI.[241] The final
part (τελευταίου ... ὄντος), in fact the remaining half (!) of the
treatise, is about the stars;[242] the sphere of the fixed stars (*4*) has to be
dealt with first (προτάσσαιτο) in a part which corresponds to books
VII-VIII, and the planets (*5*) will be treated in a part which corresponds to books IX-XIII. A complicated division, or rather blend of
divisions: a bipartite diaeresis of the general versus the particular,
the particular being next divided dichotomically into the easier
and the more complicated; a tripartite division according to
subjects, viz. (*a*) 1.3–II the end, (*b*) III-VI, and (*c*) VII-XIII. (*a*), (*b*)
and (*c*) moreover are each again being divided into two, and (*b*) is
almost twice as big as (*a*), just as (*c*) is twice as big as (*a*) and (*b*)
together.. The quantitative aspect of this tripartite division is to some

---

[239] Same division in the *Apotelesmatica*, see next section, and in the
Geographia, see the πρόλογος of book II (2.1, 1.61.3 ff. Nobbe): τὰ καθόλου have
now been treated, viz. in book I (contents briefly summarized), and 'from
here (ἐντεῦθεν) we shall begin with the exposition κατὰ μέρος'. A concise and
systematic listing of the contents of *Geogr.* books II-VII follows; we note that
books III-VII do not have proems, presumably because they do not need to.
Only book VIII has again an introduction (2.192.5 ff. Nobbe), in which
Ptolemy says that the geographical exposition is now complete, and that all
that remains to be added are the maps. The heading of the 1st chapter of book
VIII is μετὰ ποίας προθέσεως (cf. below, n. 257) δεῖ ποιεῖσθαι τὴν κατὰ τοὺς πίνακας
διαίρεσιν τῆς οἰκουμένης.
[240] For the sequence easier—more complicated see above, n. 110 and text
thereto.
[241] Note that this is again announced, after the summary of books I-II
(I.1.190.15-6, τοῖς πρὸ τούτου συντεταγμένοις) in the proem to book III, I.1.191.5-6,
ἐφεξῆς τούτων τὸν περὶ τοῦ ἡλίου καὶ τῆς σελήνης ... λόγον. The proem to book IV
(included in the first ch.) states that, the sun having been dealt with ἐν τῷ πρὸ
τούτου, it now is the turn of the moon to be treated (I.1.265.9-13). Books V and
VI too lack a separate proem, though each time it is made clear that another
book is to begin (cf. above, n. 238, and see further below, n. 242).
[242] Note that the proem of book VII (for the second dedication to Syrus see
below) is again part of the first chapter; book VIII has no introduction at all,
while the introductory passages of books VIII-XIII, briefly summarizing the
contents of the previous and announcing the subject of the present book, are
part of the first chapters.

degree paralleled in Porphyry's edition of Plotinus, where *Enn.* I-III, IV-V and VI each fill a volume of our OCT editio minor; this corresponds exactly to the contents of Porphyry's three σωμάτια (*VP* 25 *init.*, 26 *init.*)—a parallel with Ptolemy which almost looks too good to be entirely coincidental. On the other hand the bipartion (*a*) + (*b*) versus (*c*) seems to be the most important for Ptolemy, since at the beginning of book VII he addresses his dedicatee Syrus again (I.2.2.4). So Heiberg's edition of the *Suntaxis* in two volumes of about equal size exactly mirrors Ptolemy's main division. This is not contradicted by the fact that Syrus is apostrophized for the third time in the Ἐπίλογος τῆς συντάξεως (I.2.608.3),[243] which briefly and with a kind of modest satisfaction recalls what had been announced in the prologue to the *Suntaxis*: a nice instance of *Ringkomposition*.

IX 2 *The* Apotelesmatica

The *Tetrabiblos*,[244] as it came to be called (think of Robbins' Loeb edition), or rather *Apotelesmatica*,[245] is an astrological work which according to the proem is a sort of pendant to the *Suntaxis*.[246] This too is a very systematic and well-organized treatise.[247] Its long introduction, which in a most interesting way conforms to a Middle Platonist pattern outlined by Albinus—a fact that, to the best of my knowledge, has not been noticed[248]—consists of three chapters: the proem with its *definition(s)*, a chapter explaining the

---

[243] It is hard to believe that this heading (actually a rhetorical *terminus technicus*), coming after a chapter which contains only tables and before a conclusion where Syrus is addressed again, is entirely unoriginal (see below, pp. 128-9, complementary note 225); perhaps Ptolemy only wrote Ἐπίλογος.

[244] Ed. Boll and Boer (1940), Robbins (1940). Note that 3.1-4 Boll and Boer correspond to 3.1-3 Robbins; the latter combines the proem and ch. 2, while the former insert the number β′ and a chapter-heading at 107.7. I shall follow the numbering of the *Teubneriana*.

[245] For the title see below, Appendix 1.

[246] The προοίμιον tells us that both astronomy and what we call astrology are concerned with the study of the heavenly bodies and with forecasting; the latter is weaker because it deals with the unstable world below the moon, and deals with the generally accepted and practised forecasting of the weather etc. and the prediction of the fortunes of individuals.

[247] Good overview of contents in Boll (1894) 118-24; for the astronomical contents see Neugebauer (1975) 2.896-900. Useful appraisal in Taub (1993) 129-33.

[248] No reference in Taub (1993).

limitations of astrology and of (pseudo-)astrologers, but strongly
defending its *possibility* entailing its status as a scientific discipline,
against its detractors with the help of arguments of mostly Stoic
provenance,[249] and a third chapter concerned with its *utility* (ὅτι καὶ
ὠφέλιμος),[250] which also contains (adapted) Stoic ingredients, e.g.
that it is useful for one's tranquillity of mind to prepare beforehand
what may be going to happen to one.[251]

The proem begins with a remark about the predictive *aim*
(προγνώστικον τέλος, 2.16, cf. 3.21) of astronomy, and then states
that this is reached in two ways, viz. one that is first both in *order*
and potency, 2.18-9, ἑνὸς μὲν τοῦ πρώτου καὶ τάξει καὶ δυνάμει (i.e.
what we would call astronomy), and one that is second, 2.31-3.2,
3.6. The first, which is to be studied for its own sake, has its own
theory (θεωρίαν) which has been expounded in its own treatise
(*scil.*, the *Megalè Suntaxis*). In the present work an account of the
second (δευτέρου), less self-sufficient and less reliable discipline
will be provided in the proper philosophical way and by *aiming* at
the kind of truth (φιλαλήθει μάλιστα χρώμενος σκοπῷ, 3.7-8) that is
within reach. It is indeed clear that Ptolemy is concerned with the
respective *aims* of the two astronomical disciplines, with their
affinity but also with what distinguishes them, and that the *order*

---

[249] A number of Ptolemy's arguments can be paralleled from other and
earlier authors, but the argument of Boll (1894) 136-55 that Posidonius is *the*
source goes too far.

[250] For this order definition/possibility/utility and Ptolemy's exposition
in these chapters as a whole cf. Albinus *Prol.* 147.7-10 Hermann: ἀρέσκει τε τῷ
φιλοσόφῳ [*scil.*, Plato] περὶ πάντος οὑτινοσοῦν τὴν σκέψιν ποιούμενον [1] τὴν οὐσίαν
τοῦ πράγματος ἐξετάζειν, ἔπειτα [2] τί τοῦτο δύναται καὶ τί μή, [3] πρὸς ὅ τί τε χρήσιμον
πέφυκε καὶ πρὸς ὃ μή. The Platonic proof-text presumably is *Phdr.* 237cd, but
Albinus' statement is an astonishing overstatement. The passage from the
Prologos is quoted by Schissel von Fleschenberg (1930) 1761 (cf. above, n. 202),
who misapplies it to Marinus' Commentary on the *Data*, from which the
sissue of the δυνατόν is absent. I have not found other parallels, though
[Longinus] *Subl.* 1.1 comes rather close: he mentions in succession the 'what
it is' and what we may call its 'possibility' (εἴγ' ἐπὶ πάσης τεχνολογίας δυεῖν
ἀπαιτουμένων, προτέρου μὲν τοῦ δεῖξαι τί τὸ ὑποκείμενον, δευτέρου δὲ τῇ τάξει, τῇ
δυνάμει δὲ κυριωτέρου, πῶς ἂν ἡμῖν αὐτὸ τοῦτο καὶ δι' ὧν τινων μεθόδων κτητὸν
γένοιτο), while a few lines before he had mentioned utility (ὠφέλειαν). On
the links of the *De sublimitate* with Middle Platonism see Donini (1969) and
(1982) 135-7.

[251] E.g. Posid. Fr. 165.28-32 Edelstein-Kidd *ap.* Gal. *PHP* 4.7.7, p. 282.10-4
De Lacy, προενδημεῖν ... τοῖς πράγμασι κτλ., see Kidd (1988) 2.601. The difference
is that in an astrological context one *knows* beforehand what *is* going to
happen. I note in passing that Hephaestion of Thebes (see below) only contains excerpts from Ptol. chs. 1.1 and 1.3.

he has in mind is in the first place *systematic*, but also has a *didactic* aspect. Clearly one can only practise what we would call astrology in a responsible way when aware of its limitations as compared with astronomy, a discipline of which moreover one needs to have sufficient knowledge precisely in order to understand why astrology comes second and how it is possible nevertheless.

The headings of the chapters 1-3 are perfectly in accordance with their contents;[252] moreover chs. 2 and 3 are announced at the end of the proem: the χρήσιμον of astrology will be treated (= ch. 3), but first its 'possibility' (πρῶτον τοῦ δυνατοῦ, 3.24-5 Boll and Boer).[253] At the beginning of ch. 3 moreover the author says that the topic of the δυνατόν has now been dealt with. We again notice Ptolemy's concern for *orderly* and *systematic* treatment (τάξις). This is also clear from the end of this chapter, 17.5-10, which briefly lists the *contents* of the following chapters of book I: he will begin with the individual character of each of the heavenly bodies and their active powers, and first discuss the sun, the moon and the other planets, in this order (the same as in the *Suntaxis*). He also states what will be his *manner of presentation*: this will be by way of an introduction (κατὰ τὸν εἰσαγωγικὸν τρόπον). And he tells us that he deals with these matters 'in the physical way', κατὰ τὸν φυσικὸν τρόπον (cf. 58.13, φυσικὸν λόγον). To understand what he means we must recall the introduction to the *Suntaxis*:[254] physics is insecure inasmuch as it is involved with matter, and it should be helped out and shored up by the use of mathematics.[255]

---

[252] For the issue involved see below, pp. 128-9, complementary note 225.

[253] See above, n. 250.

[254] Above, text to n. 229. For the εἰσαγωγικὸς τρόπος of Nicomachus see below, Ch. XI 1; the formula is not often found: parallels at Did. Caec. *in Gen.* cod. 114.4, Ammon. *in Isag.* 47.3 Busse, Elias *in Isag.* 44.6 Busse (for the equivalent formula ἐν εἰσαγωγῆς τρόπῳ see Porph. *Isag.* 1.8 Busse and his commentators *ad loc.*, Iambl. *VP* ind. cap. 17.3, τοῦ τρόπου πρὸ τῆς εἰς φιλοσοφίαν εἰσαγωγῆς, Eus. *Gen. elem. introd.* 3.13-4 Gaisford, [Gal.] *Philos. hist.* 24.3). For examples of works with the word εἰσαγωγή in the title see e.g. De Libera and Segonds (1998) 31. φυσικὸς τρόπος in the sense meant by Ptolemy is equally rare: Ascl. *in Met.* 136.18 Hayduck, Dam. *in Phaed.* 123.7 Westerink, Philop. *in Phys.* 57.11 Vitelli.

[255] Quite similarly, in the introduction to the *Harmonica* (1.1-2) he argues that harmonics (or canonics, as it is also called) is both theoretical and involved with imprecise sense-perception, and that the best way to treat the subject is to adjust the data of acoustics with the aid of reason, which is superior. The Pythagoreans are too theoretical where numbers in relation to the world of sense-perception are concerned, while the Aristoxeneans are not theoretical enough.

Book II is concerned with general matters, that is to say with the major and minor events that will befall whole peoples, countries, cities. This is the so-called καθολικόν part—a term also found in the title of Hephaestion of Thebes' book I, which contains a number of extracts from Ptol. *Apotel.* I-II; also cf. the proem to his second book, 61.4-5 Pingree. The first chapter of Ptol. *Apotel.* II in a number of mss. is not unaptly entitled '*division* (*scil.*, into four subparts)[256] of the general investigation', διαίρεσις τῆς καθολικῆς ἐπισκέψεως. For, as Ptolemy says, astronomical prognostication is *divided* into *two parts*, the general part and the so-called genethlialogical part which pertains to the horoscopes of individual humans. The general part will be treated first.

Book III has again a προοίμιον, in which books I-II are summarized and prognostication concerning humans announced. Since here the moment of conception (more difficult to establish however) and that of birth are most important, these will be the first to be discussed (viz. in ch. 2). In this chapter Ptolemy also looks back at the second chapter of book I (see above) which he calls 'the ἐπιλογισμός ('consideration', 'reflection', 'argumentation') at the beginning of the present treatise', and states that in the present *section* (μέρος) too it is his (authorial) *intention* (προθέσεως)[257] to avoid the complicated practices and the mistakes of the astrological dilettanti. The next topic (treated in ch. 3) will follow according to the proper *systematic ordering*, κατὰ τὴν προσήκουσαν τῆς τάξεως ἀκολουθίαν (110.5). In ch. 4 the contents of the rest of book III and of the whole of book IV are listed meticulously topic by topic under the apposite heading διαίρεσις γενεθλιαλογίας—again a *division into parts* according to a *systematic ordering*.[258] We may move quickly to the concluding section of the final chapter of book IV, extant in a

---

[256] Listed 57.18-58.2.

[257] The (apposite) chapter heading of *Harm.* 1.2 is τίς πρόθεσις ἁρμονικοῦ (defined 4.13-5 Düring)—here πρόθεσις is generalized and becomes the aim of the professional, but this professional is of course and in the first place Ptolemy himself. For this pseudo-generalization cf. Ptol. *Geogr.* 1.2, 5.17-20 Nobbe, τί μὲν οὖν τέλος [see below, pp. 122-3, complementary note 11] ἐστὶ τῷ γεωγραφήσοντι ... ὑποτετυπώσθω.

[258] εἴ τις αὐτῆς τῆς τάξεως ἕνεκεν διαιροίη τὸ καθ' ὅλου τῆς γενεθλιαλογικῆς θεωρίας, 112.14-5; for τάξις see also 113.14 and 115.11. The second book of Hephaestion, containing a number of extracts from Ptol. *Apotel.* III-IV, has the word γενεθλιαλογικόν in its title according to the *pinax* (and it is supplemented in the text of the treatise by Pingree).

single ms. only,[259] where according to the longer version Ptolemy says that he has now fulfilled the πρόθεσις which has been stated at the beginning, *scil.* of his treatise.

It will be clear that Ptolemy in this work too uses what came to be systematized as preliminary isagogical questions, including technical vocabulary, though he does so in a free and unpedantic way, just as was the case in the *Suntaxis*.

---

[259] Printed not in the text but *in app.* in the *Teubneriana*, but convincingly defended by Robbins against that of the epitome ascribed to Proclus, for which see below, text to n. 284. This other version has σκόπος at 213.2 Boll and Boer.

CHAPTER TEN

COMMENTARIES ON PTOLEMY

X 1 *Pappus and Theon on the* Mathèmatikè Suntaxis *and* Handy Tables[260]

We still have Pappus' Commentary on books V-VI of the *Suntaxis*, and (incomplete) Theon's Commentary on books I-XIII (book III was revised by 'my daughter Hypatia').[261] In the parts that are extant Pappus refers to his Commentaries on books I (255.1 Rome) and IV (76.20-1 Rome).[262] It is likely enough that his *commentarius perpetuus* also included books II and III,[263] possibly even the whole work.[264] Explicit backward references such as those just cited decidedly convey the impression that what we have here are the remains of an authorized publication.

Pappus' introductions to each of these books are no more than

---

[260]   See below, p. 129, complementary note 260.
[261]   Ed. Rome: Pappus V-VI (1931), Theon I-IV (1936) and (1943); on Hypatia's role in Theon's third book see Knorr [1989] 754-63). Note that book XI of Theon is lost and that of book V only a fragment (see Rome [1953]) is extant. The other books have not yet found a modern editor (Rome's collations were destroyed), so the only available edition (*non vidi*) of the subsequent books is still that in Grynaeus and Camerarius (1538); see Tihon (1978) 1-2, and Toomer (1976b) 321-2, 324, who also dwells on Theon's use of Pappus. An anonymous Commentary on the *Handy Tables* (see above, n. 222) contains a number of references to Pappus, possibly to the Commentary on the *Suntaxis*, see Tihon (1978) 171-83, though perhaps it is not to be excluded that Pappus also commented on the *Handy Tables*. The *Handy Tables* (*Procheiroi Canones*, see above, n. 223) is a handbook for astrologers, mostly consisting of astronomical tables.
[262]   See Rome's notes *ad locc.* (1931) 255-6, 76.
[263]   For a reference in an Arabic Commentary to book III see Neugebauer (1975) 2.966.
[264]   Ziegler (1949)1087-8 is in my view hypercritical. He bases his view that Pappus commented on only part of the *Suntaxis* on the *Suda* lemma on Pappus (Π 265, 4.26.6 Adler), where a title of Pappus is formulated as εἰς τὰ δ' βιβλία τῆς Πτολεμαίου Μεγάλης συντάξεως ὑπόμνημα. There must be some mistake here, perhaps through *saut du même au même* and *Verschlimmbesserung*: read e.g. εἰς τὰ δ' βιβλία [τῆς] Πτολεμαίου ⟨ὑπόμνημα, εἰς τὰ ιγ' βιβλία τῆς Πτολεμαίου⟩ Μεγάλης συντάξεως ὑπόμνημα. If this speculation is correct, Pappus would also have written a Commentary on the *Apotelesmatica*. But more probably the number in the *Suda* is simply wrong.

extremely detailed summaries of their contents: 'Ptolemy in book V' treats the following, chapter by numbered chapter, 'in book VI' the following, again chapter by numbered chapter. Accordingly Pappus is concerned with the *theme(s)* and the meticulously precise *division into parts* of these books. At 173.24 Rome he tells us that he has given his *summary* of book VI for *didactic* reasons (ταῦτα ... ὡς ἐν περιοχῆς λόγῳ ὑπομνήσεως ἕνεκεν εἴρηται). What is of course also clear (though he does not say so in so many words) is that he has no doubts that the *Suntaxis* is correctly ascribed to Ptolemy, i.e. is *authentic.* So with some effort we are in a position to show that several isagogical issues are applied. Still, Pappus is far less clear about these matters than in his Commentary on *Elements* X, or in the *Collectio.* Perhaps more was to be found in his lost introduction to the whole work and to book I, perhaps not; we just don't know.

Theon's Commentary on the *Suntaxis,* as he says himself at the beginning, was composed and published at the request of his *students* (317.2-18.21 Rome). In his general introduction (the part which interests us in the present context), which at the same time is a commentary on Ptolemy's proem (i.e. ch. 1 of *Synt.* book I) he complains that his predecessors in their Commentaries have skipped things that were difficult, or omitted to provide mathematical proofs, so he was obliged to add a lot himself (318.5-9). One wonders whether he includes Pappus whom he followed to a degree. Perhaps the remark is to some extent merely a hackneyed *topos.*[265] He also tells us that he will deal with book I κατὰ λέξιν (319.23), which by the way is far from true, and more succinctly with the others. But difficulties will be explained (318.14), even in the later books (319.4). Ptolemy's προοίμιον (as he calls it *disertis verbis,* 319.6, cf. 324.12-25.1) is *clear* enough (σαφές) and intended for the *young* (τοὺς νέους)—a remark pertaining to the *manner of presentation* and to the *qualities* to be expected of Ptolemy's *students.* Theon indulges in quite an amount of simple paraphrase and elementary elucidation. What is interesting is that he confirms the headings of the first three chapters; the proem has already been mentioned, and Theon's second and third chapters have the same headings as the corresponding chapters in Ptolemy.[266] At 334.9-10 he even explains the *reason why* Ptolemy gave its *title* to

---

[265] For this *topos* cf. above, n. 117. See Toomer (1976b) 321, who remarks that Theon's "trivial exposition" may be criticized on the same grounds.

[266] For the issue involved cf. below, pp. 128-9, complementary note 225.

ch. 3 (ἀπὸ ταύτης καὶ τὴν ἐπιγραφὴν τοῦ κεφαλαίου πεποιῆσθαι.) Not surprisingly he is also concerned with the *systematic and didactic order* of Ptolemy's work (327.1-2, ἀκόλουθον ποιεῖ καὶ τὴν τάξιν τῆς τε τούτων διδασκαλίας, cf. 330.19-20), even calling Ptolemy's presentation of his general and particular topics τὴν ἀπαρίθμησιν τῶν τε καθόλου καὶ κατὰ μέρος (334.2). *Utility* is not forgotten either (τὸ χρήσιμον).

It is clear that Theon knows what isagogical questions are, but also that he employs them in a rather off-hand way. Even so, he is merely following in Ptolemy's footsteps. However nothing of the kind is to be found in the introduction to the commentary on book II of the *Suntaxis*, which merely summarizes the contents of the previous book and tells us what to expect in this one. The same holds for the commentaries on books III and IV.

After his Commentary on the *Suntaxis* Theon wrote two works on the *Handy Tables*: a substantial treatise in five books, next a short tract in one.[267]

The *Great Commentary*,[268] as it is commonly called (though it is not a Commentary in the proper sense of the word), has little to offer in our present context. The proem of book I (there is good ms. evidence for the heading προοίμιον here), dedicated to two pupils which seems to show that it was intended to be formally published, is quite short. The only remark of any interest is that this treatise is intended for those who have made a certain progress in mathematics[269] in conformity with (Ptolemy's) *Suntaxis*, so has to do with an *order of study* and with the *qualities expected of students*.

The *Little Commentary*, as it is commonly called, though it is not a Commentary either but a number of sets of untechnical instructions distributed over chapters,[270] also has little to offer for our present purpose. True, there is an introductory chapter defining the terminology which has to be taught and expounded *before* one can go on (προδιδάξαι 200.9, προδιειλημμένων 202.1 Tihon). Next one is

---

[267] Books I and II-III of the *Great Commentary* have been edited by Mogenet and Tihon (1985) and Tihon (1991), so book IV is not yet available (book V is lost); the *Little Commentary* has been edited by Tihon (1978).

[268] On the state of the text of book I (draft (?), transmission, revision(s)) see Mogenet and Tihon (1985) 69-80.

[269] This echoes a remark of Ptolemy, see above, n. 234 and text thereto, and Mogenet and Tihon (1985) 158 n. 2.

[270] The title in Tihon (1978) is Θέωνος Ἀλεξανδρέως εἰς τοὺς προχείρους κανόνας. Some mss. add ἑρμηνεία or παράδοσις (the latter also once in the explicit).

told about the technical details which have to be learnt *before* the rest (προμανθάνειν, 202.3). Perhaps more interesting is Theon's remark at the beginning that this tract about the *Handy Tables* is meant for those who do not have a sufficient knowledge of arithmetic and are entirely unacquainted with geometric proofs (199.3-10). This is an implicit definition of the *aim* of the Little Commentary, and involves the *qualities* (or rather lack of them) to be expected *of the student.*

X 2 *The Anonymous Introduction to the* Mathèmatikè Suntaxis

A number of mss. contain an introduction to Ptolemy combined with a commentary on selected passages of Ptolemy's *Suntaxis* book I, which has not yet been published in its entirety.[271] Mogenet attributed the work to Eutocius, but this attribution has been refuted by Knorr.[272] What we have here is a compilation based on a plurality of sources, among whom Pappus and Theon; mention of the philosopher Syrianus provides a *t.p.q.*,[273] and shows, or so I believe, that the author was a member of a Neoplatonist establishment. In our present context the first section,[274] the Προλεγόμενα as they are called in several mss., is of major interest, though we may note in passing that the treatise as a whole conforms to the division 'before the work', *ante opus* (the prolegomena), and 'on the work itself', *in ipso opere* (the commentary proper, σχόλια).[275] Mogenet in his pioneering study of 1950 proved that this tract must be late because this section in a scholastic and explicit way, and using the full technical vocabulary, deals with the following six isagogical issues: (1) the σκοπός,[276] i.e. providing irrefutable geo-

---

[271] The edition prepared and promised by Mogenet has not appeared. For editions of parts of the work see Mogenet (1956) 6-8; further above, text to n. 49, below n. 274.

[272] Mogenet (1956) 12-34, Knorr (1989) 155-211. The main point is that the section on isoperimetric figures (attributed to Eutocius by Mogenet on the basis of two other texts on this topic by Eutocius) cannot be by Eutocius; Knorr *ibid.* 161 caps this with a linguistic argument. See further below.

[273] Mogenet (1956) 9.

[274] Published by Hultsch (1876-8) 3.xvii-xix, who attributes it to Pappus on the basis of a guess in a late ms.

[275] Cf. above, n. 201 and text thereto.

[276] Mogenet (1956) 19 shows that Hultsch's reading σκοπόν is wrong. In my view it follows that the preceding words ἥντινα σύνθεσιν should be bracketed, or daggered.

metric proofs for the (astronomical) phenomena; (2) the χρήσιμον, which follows from the fact that it is beyond sectarian partiality; (3)/(4) the τάξις[277] and τὸ γνήσιον, which are self-evident; (5) the εἰς τὰ μόρια διαίρεσις, set out at length 3.xviii.17-xix.18 Hultsch, first in general terms and then as to the contents of the individual books; and finally (6) the explanation of the title: 'it is entitled *Suntaxis* because the bare and unproven approaches of the *Handy Tables* are systematically linked with each other by logical and linear [or rather: geometric] demonstrations' (ἐπιγέγραπται δὲ Σύνταξις διὰ τὸ συντετάχθαι ταῖς λογικαῖς καὶ γραμμικαῖς ἀποδείξεσι τὰς τῶν Προχείρων κανόνων ψιλὰς καὶ ἀναποδείκτους ἐφόδους).[278]

Mogenet failed to notice that the very first section (3.xvii.5-19 Hultsch) of these prolegomena, which defines astronomy by quoting the definition from ch. 1 of Ptolemy's *Apotelesmatica*,[279] and then explains the terms of this definition, in fact tells us *to what part, or section*, of mathematics this particular discipline belongs: another isagogical question, which however became an ingredient of the explicit scholastic scheme a bit later than the others.[280] It is not formulated explicitly here either but its actual presence is undeniable, and the prominent position awarded to it, viz. at the very beginning of the exposition, suggests that it is important to the author of this piece. Presumably his model for the introduction as a whole is the familiar division *de arte, de opifice, de opere*,[281] though in the present case the section *de opifice* is lacking.

We have noticed above that the mathematical argument provided by Mogenet for attribution to Eutocius has been refuted by Knorr, who argues that the author is an otherwise unkown person called Arcadius, mentioned by Eutocius as a commentator on Ptolemy.[282] But the fact that the section on isoperimetric figures is

---

[277] Note that *Synt.* 1 ch. 2 is appositely entitled Περὶ τῆς τάξεως τῶν θεωρημάτων.

[278] This echoes a remark of Theon *in Synt.* 318.11-2 Rome on other commentators on this work: 'for the most part they draw their conclusions, as in *Handy Tables*, by means of unsupported arguments', αὐτοὶ τὰ πλεῖστα καθάπερ ἐν προχείροις κανόσι διὰ ψιλῶν ἐφόδων περαίνουσιν. 'Linear demonstrations' are proofs according to the *mos geometricus*, see e.g. Hintikka and Remes (1974) 99.

[279] *Apotel.* 1.1, p. 2.16-21 Boll and Boer. The *Suntaxis* fails to provide such a brief and handy definition, so Mogenet's scorn regarding Anonymus' taking the definition of astronomy from the astrological work is a bit unfair.

[280] Mansfeld (1994) 11, 15, 19.

[281] See above, n. 210 and text thereto.

[282] *in Arch. De sphaer. et cyl.* 3.120.8; see Knorr (1989) 165-6.

not by Eutocius and the further fact that Eutocius read an exposition concerning isoperimetric figures in Arcadius does not prove that this section in the Anonymus is by Arcadius, but only suggests that it could be. However this may be, that the commentary—cum—introduction cannot be by Eutocius is confirmed by the fact that the author uses the explicit scholastic scheme of the isagogical issues in a matter-of-course and, so to speak, almost tired way. We have seen above that Eutocius' own procedure in the genuine Commentaries is quite different.

It is not to be precluded that these *Prolegomena* to some extent go back to and are a systematization of Pappus' introduction to the first book of the *Suntaxis* in the lost first book of his Commentary, but this remains entirely speculative.

X 3 *Commentaries on the* Apotelesmatica

Several Commentaries and comments on this work survive,[283] but these are of little use for the present enquiry. The Paraphrase of the *Apotelesmatica* ascribed to Proclus (of which no critical edition exists)[284] is probably inauthentic, and it is anyhow nothing but a relatively short and (in respect of our purposes) uninformative paraphrase. The *Eisagoge* by or ascribed to Porphyry[285] is from our point of view equally disappointing; the only remark of some interest is found in the προοίμιον: it is the author's purpose to explain Ptolemy's difficult and old-fashioned terminology for the sake of *clarity*, σαφηνείας ἕνεκεν (190.8-10 Boer and Weinstock).[286]

---

[283] See Gundel and Gundel (1966) 213-6.
[284] I have seen Allatius (1731).
[285] Ed. Boer and Weinstock (1940).
[286] Quoted Mansfeld (1994) 204.

CHAPTER ELEVEN

NICOMACHUS OF GERASA AND HIS COMMENTATORS

XI 1 *The* Introductio Arithmetica

This popular and influential treatise[287] about arithmetic, or rather
the theory of numbers, has been briefly mentioned above.[288] It
consist of two books and is structured very clearly. The prologue
(1.1-5) does not lay much explicit emphasis on isagogical questions, because the *Introductio* as a whole is isagogic, that is to say
prepares the way for the *Theologoumena* (as we know though
Nicomachus here does not tell us).[289] Again and again he insists
on the *manner of presentation* of arithmetic in the present treatise:
this is no more than a (preliminary) *introduction*, εἰσαγωγή: 1.19.20
at 55.4 Hoche; 1.22.4 at 64.22-3, 2.12.1 at 95.14, and 2.29.5 at 147.1-2,
final sentence of the treatise, ὡς ἐν πρώτῃ ... εἰσαγωγῇ; full and
accordingly genuine *title* 2.22.3 at 123.15-6, αὐτὴν τὴν Ἀριθμητικὴν
εἰσαγωγὴν πρὸ πασῶν τῶν ἄλλων ὑπάρχειν. Another term used by
him is τεχνολογία, 'systematic treatment',[290] viz. of these indispensable *introductory* matters. At the end of his prologue he states (1.5 at
11.20-4):

> So then we have rightly undertaken *first* the systematic treatment
> (πρότεραν τὴν τεχνολογίαν) of this [*scil.*, preliminary arithmetic], as
> the science naturally prior, more honourable, and more venerable,
> and, as it were, mother and nurse [of the other mathematical
> disciplines],[291] and for the sake of *clarity* (τοῦ σαφοῦς χάριν) we

---

[287] Overview of Commentaries and revised versions (Iamblichus, Boethius) at D'Ooge *& al.* (1926) 125-32. On Nicomachus see Tarán (1974), Dillon
(1977) 352-61, Donini (1982) 140, Hadot (1984) 63-9, O'Meara (1989) 14-23,
Dörrie and Baltes (1993) 68-71, 269-71 (also for further references to the
literature).

[288] Text to nn. 58 and 59.

[289] Above, text to n. 58. On the structure of technical handbooks in general
see Fuhrmann (1960), where however Nicomachus is lacking.

[290] Forms of the verb (τεχνολογεῖν) are first found in Aristotle's *Rhetoric,*
and in the first two chapters of this treatise only in the whole of the *Corpus
aristotelicum*; here they pertain to the authors of rhetorical *technai* (1.1.1354b17,
b27, 1355a19, 1.2.1356a16). The verb, and the noun τεχνολογία are later also
applied to other disciplines.

[291] Cf. 1.4.1 at 9.8: 'origin, root, and mother'.

shall make our beginning of this systematic treatment from here onwards.[292]

Compare 1.17.1 at 44.8-10: 'Now that we have given a *preliminary* systematic account (προτεχνολογουμένου) of absolute quantity, we shall turn to relative quantity'. This use of the term προτεχνολογου-μένου is not that of designating an isagogical scheme;[293] it only refers back to a particular section of the treatise. But the treatise as a whole may be seen as the Προτεχνολογουμένα of Arithmetic with a capital Π and A.

Even so, the prologue of the treatise deals with issues that may be termed isagogical; more specifically, with its *theme*, viz. arithmetic, defined and described at some length, and with the status of this science vis-a-vis the other subdisciplines of mathematics, all of which are dependent on it. So the isagogical issues of *utility*, of the *systematic sequence* and *order of study*,[294] and of the ὑπὸ ποῖον μέρος ... ἀνάγεται are all co-involved (cf. the concluding section of ch. 1.5, quoted a moment ago). In 1.3.1-3 at 5.13-6.8 Nicomachus is quite specific about the relations between arithmetic, geometry, 'music' (i.e. canonics), and 'spherics' (i.e. astronomy): arithmetic is prior to canonics, and geometry to astronomy.[295] The *utility* of the mathematical sciences for human life (εὔχρηστά εἰσι πρὸς τὸν ἀνθρώπινον βίον, 8.10-1)[296] is illustrated by means of an exegetic paraphrase and partial quotation of Plato, *Resp.* 7.522c ff.:[297] arithmetic is useful for distributions, etc., geometry for the founding of cities, etc., 'music' for festivals, etc., and astronomy for farming, navigation, etc. (1.3-7 at 8.8-9.4).

Chapters four and five deal with the *systematic* and didactic sequence, i.e. *order of study*, of these four disciplines: 'which is the

---

[292]   Transl. D'Ooge, modified; my italics.
[293]   Examples above, n. 187; for Heron's more technical use which more-over is earlier see above, text to n. 183. For the term at *Ar.* 2.6.1 see text to n. 302 below.
[294]   On this order of study cf. Hadot (1984) 67-8.
[295]   That canonics (or 'harmonics' as he calls it) is subordinated to arithmetic is also Aristotle's view, *APo* 1.13.78b38. For Aristotle on the relations between the various pure, applied and empirical mathematical subdisciplines see Ross (1949) 554-5, Barnes (1975) 151-5, Detel (1993) 2.301-9. For Theon of Smyrna's division, similar to Nicomachus', see *Util.* 16.24 ff. Iamblichus' sequence is more conventional than Nicomachus': arithmetic, geometry, canonics, astronomy (?), see pinax of his treatise *On Pythagoreanism* at O'Meara (1989) 31-5.
[296]   Cf. above, n. 71.
[297]   Cf. above, n. 12.

84 CHAPTER ELEVEN

first that must be learned?' (τίνα οὖν ἀναγκαῖον πρωτίστην ...
ἐκμανθάνειν, 9.5-6). The answer is unambiguous: arithmetic comes
first, not only because it is first in the divine mind, but also because
it is not destroyed if the others are abolished, while all the others
vanish if arithmetic is done away with. So it is absolutely primary.
In the first place arithmetic comes before geometry. Secondly, it
comes before music. Finally, also astronomy is entirely and ultimately dependent on it; indirectly, since it depends on geometry
and 'music' which are tributary to arithmetic themselves, but also
directly, because the various forms of behaviour of the heavenly
bodies are determined by numbers. The final section of ch. 5,
which is about the *clarity* which determines the exposition of what
comes next, has already been quoted.

The last chapter of book I (1.23.4 ff. at 61.24 ff.) is also interesting,
because another and even more important arithmetical approach is
introduced and then explained in some detail, which is 'more
subtle and most *necessary* (ἀναγκαιοτάτη—issue of *utility* again) for
the physical study of the universe'. This method shows to us in a
way which is at the same time absolutely *clear* (σαφέστατα) and
irrefutable, that what is beautiful, limited, and knowable (ὑπὸ
ἐπιστήμην πῖπτον) is prior to its opposite, and that the parts and
species of this opposite are given shape, limit, *order* and proper
sequence by what is beautiful, limited, and knowable. They are so to
speak 'stamped' by it (65.7). All the species and specific differences
of inequality are determined and produced by equality (65.18-21).

Book II of the *Introductio* too contains a number of passages that
are of interest in our present context. A general method (74.15) is
introduced, which has as its corollary a theorem that is extremely
useful for understanding the Platonic psychogony[298] (χρησιμώτατον
εἰς ... τὴν πλατωνικὴν ψυχογονίαν, 76.14-6), as well as for for understanding harmonic intervals in general. This is demonstrated in
chapters three and four. At the beginning of ch. 5 Nicomachus
says: 'We have made *clear* (σαφηνίσαντες) what further ratios are
produced by combining ratios; what is left is to proceed with what
follows of the Introduction' (80.1-3). Note the emphasis on the
ordered exposition.

---

[298] *Tim.* 35a ff. Cf. *Ar.* 2.24.6 at 129.16-7, χρησιμεύοντος ἡμῖν εἰς Πλατωνικόν τι
θεώρημα.

The first section of chapter six of book II too deserves to be quoted:[299]

> We now have sufficiently expounded relative quantity, by a process of selection measuring out what is appropriate and easily comprehensible for *beginning students* (τῇ τῶν ἄρτι εἰσαγομένων ἕξει).[300] Whatever remains to be discussed about this part (τόπον)[301] will only be supplied after we have put it aside, and first given a *preliminary systematic exposition* (προτεχνολογησάντων)[302] of other things [...]. For mathematical theorems are after all articulated and *clarified* (σαφηνίζεσθαι) through each other. What is to be investigated and looked into *before* (the rest of this part) has to do with linear, plane and solid numbers [...]. Naturally, proper instruction about these numbers belongs in the *Introductio Geometrica* (ἃ δὴ ἰδίως μὲν ἐν τῇ Γεωμετρικῇ παραδίδοται εἰσαγωγῇ), as they are more related to magnitude [viz. than numbers without extension]. Even so, the germs of these entities are included in arithmetic, since this so to speak is the mother of geometry, and born before it.

Several details claim our attention. The first of these is that Nicomachus is quite explicit about the isagogical issue of the *qualities* to be expected of the *student*. He writes for beginners. The words used show that he could have called his treatise Περὶ ἀριθμητικῆς τοῖς εἰσαγομένοις, a title which of course is entirely equivalent to the title Ἐισαγωγὴ ἀριθμητική he *did* choose. One only has to recall the titles of some introductory works by Galen (all of them extant): the Περὶ αἱρέσεων τοῖς εἰσαγομένοις, the Περὶ σφυγμῶν τοῖς εἰσαγομένοις, and the Περὶ ὀστῶν τοῖς εἰσαγομένοις.[303] We also note the emphasis on *clarity* and *orderly exposition*, and the fact that the excursus on numbers with extension is *preliminary to* (cf. προτεχνολογησάντων) the proper treatment of relative quantity. Another piece of information which is not without significance is that Nicomachus refers to an *Introductio geometrica* which presumably either has already been

---

[299] 82.10-83.7; transl. D'Ooge, modified, my italics.
[300] D'Ooge mistranslates: "to the nature of the matters thus introduced". Better Bertier (1978) 101: "selon la nature des débutants". A quite common formula. Cf. Procl. *in Eucl.* 272.12-4 Friedlein, ὧν τὰς ἐπινοίας δυσθεωρήτους οὔσας τοῖς εἰσαγομένοις παραλείπομεν ἐν τῷ παρόντι, and see further below, n. 303 and text thereto.
[301] Cf. above, n. 25.
[302] Cf. above, n. 290 and text thereto.
[303] *De ord. libr.* ch. 2, 19.54 Kühn = *Scr. min.* 2.84.2-7 Mueller, *Ars. med.* 1.408.17-09.2, ὅσα τοῖς εἰσαγομένοις ἐποιησάμεθα, τὰ περὶ ὀστῶν, καὶ ἡ τῶν μυῶν ἀνατομὴ, καὶ ἡ τῶν νεύρων, καὶ ἡ τῶν ἀρτηριῶν καὶ φλεβῶν, καί τινα τοιαῦτα ἕτερα., *ibid.* 410.6-7 (numerous other instances in Galen). See Mansfeld (1994) 198, and cf. above, text to n. 56, n. 300, and below, pp. 123-4, complementary note 56.

written by him, or which he plans to write, but which in the *order of study* clearly came after the *Introductio arithmetica*, just as geometry is posterior to arithmetic.[304] Accordingly, that some things expounded in this other work are anticipated here is both unavoidable and helpful.[305]

A related passage is found at the end of chapter twenty-four, which is equally informative as to Nicomachus' practice as a teacher, and again concerned with a proper *order of study*, and with *clarity*, viz. 2.24.10 at 131.7-9: 'these matters [viz., certain multiplications] will receive their proper *clarification* (σαφηνείας) in the reading of Plato in class (ἐν τῇ Πλατονικῇ συναναγνώσει), that is to say the passage on the so-called marriage number' (i.e. *Resp.* 8.546a ff.) The expression 'in the Platonic reading in class' either refers to a lecture (or lectures) on this passage Nicomachus intends to give, or to the written account of such a course. Compare, in the 'second problem' at Plu. *Quaest. conv.* 8.2, 700C, the phrase ἐν ταῖς Πλατωνικαῖς συναναγνώσεσιν ὁ λεγόμενος 'κερασβόλος' καὶ 'ἀτεράμων' [*Leg.* 853d] ζήτησιν ἀεὶ παρεῖχεν, clearly referring to studying a Platonic text, however informally, in class and encountering on this occasion an issue now recorded in writing.[306]

We may also briefly look at the references to the 'ancients' in Nicomachus: 2.28.1 at 140.14-8, the treatment of the three proportions παρὰ τοῖς ἀρχαίοις (begun 2.22.1) is now completed; it has been set out more *clearly* (σαφέστερον) and in more general terms because it is encountered frequently though in manifold forms in the studies of their writings (ἐν τοῖς ἀναγνώσμασι). In 2.28.6 at 142.22-43.1 it becomes clear that these ancients are Aristotle and Plato as followers of Pythagoras. The word παλαιός is found more often: 1.1.1 at 1.5-6, Pythagoras and those who came after him;

---

[304] Heath suggests (1921) 1.97 that this "may not necessarily have been a work of his own". But the *Fihrist*, Dodge (1970) 2.643, attributes such a work in two books to Nicomachus. Also see O'Meara (1989) 86-7.

[305] We may for example compare Chrysippus' practice, according to which the order of study is logic—ethics—physics-cum-theology; criticized by Plu. *S.R.* 1035AF because he stated that ethics has its foundation in theology, and remarked in his Περὶ λόγου that the student who begins with logic need not keep away altogether from the 'others', viz. ethics and physics, but is to touch upon them as the circumstances require.

[306] For the important term συνανάγνωσις see Mansfeld (1994) 245, index *s.v.* reading, and above, Ch. VI 8. For suggestions as to Nicomachus' teaching of Plato see Haase (1982) 88 ff.

2.17.1 at 109.3-4, Pythagoras and his *successors* (διαδόχους);[307] 2.18.4 at 114.7-15, Plato and Philolaus; 2.21.1; 2.22.1 (see above) at 122.11-3; Pythagoras, Plato and Aristotle, 2.28.1 at 140.14-6. Still, for the sake of completeness Nicomachus sometimes also includes later developments; see 2.22.1, where we are told that first three kinds of proportion were added to the triad of the ancients, and that οἱ νεώτεροι (122.17-8) discovered four more such proportions. The older ones are treated at greater length, and in proper *order* (τάξει, e.g. 131.13). References to one's predecessors, whether critical or not are, as we have seen several times, a quite common *historical* element of the thematic ingredient of 'introductions'.

XI 2 *Iamblichus' Version and Asclepius' and Philoponus' Commentaries*

Iamblichus' *in Nicomachi Arithemeticam introductionem liber*,[308] as already said in Ch. I 1 above, is not a Commentary. It is a clearly written and free paraphrase of Nicomachus' treatise, interlarded with extra material such as quotations from purported Pythagorean authors. In his proem Iamblichus states what on all accounts is the *aim* of this section of his multi-volume work on Pythagorean philosophy, viz. to treat arithmetic, the primary mathematical science. But he almost immediately adds that everything one needs is found in Nicomachus' Ἀριθμητικὴ τέχνη (4.12-4 Pistelli). No information on Nicomachus himself however is provided,[309] apart from a eulogy of his capabilities and the qualities of his exposition (4.14 ff.); its *systematic order* is singled out for special praise (τάξιν θαυμαστήν, 4.17-8). Otherwise, there is little to interest us in our present context. In fact, Iamblichus is much less scholastic than Nicomachus, at least in the present work.

Some Commentaries on this work mentioned in our ancient sources are lost,[310] while others are extant. Tarán in his exemplary

---

[307] See further below, Appendix 2. In Mansfeld (1992) I should have paid attention to the fact that this constructed Pythagorean succession, which is expounded at length in Hippolytus' *Refutatio* and in fact forms the basis of his attack against the Gnostics is explicitly attested in Nicomachus. But note that Hippolytus included Empedocles, Heraclitus and the Stoics as well.

[308] Ed. Pistelli (1894). See below, p. 130, complementary note 308.

[309] From the letter of dedication and epilogue to his *Harmonica* (237-8, 265 Von Jan) we know that Nicomachus travelled around a lot and so only was able to write a short introductory vademecum (ἐγχειρίδιον) on this subject. This, at least, is what he claims. In general see Haase (1982), esp. 120 ff., 159 ff.

[310] The Commentary rightly or wrongly ascribed to the hierophant

monograph has argued convincingly that the closely related Commentaries of Asclepius and Philoponus,[311] both pupils of Ammonius Hermiae, either derive from a shared set of notes of a course given by Ammonius or, more probably, that Philoponus "edited" Asclepius' version or a version very close to that of Asclepius.[312] Both refer to Ammonius as 'our master'.[313] These rather thin Commentaries do not provide much information that is of interest in our present context. Even so, there are a few titbits worth looking at.

Both Asclepius and Philoponus in their comment on the first lemma tell us that the author is a Platonist (not a Pythagorean!)[314] and pursues a Platonic *aim*, Πλατωνικὸν σκοπόν, viz. the τέλος of real philosophy plus the road which leads towards this goal (via arithmetic and then the other mathematical sciences, of course).[315] That this is the σκοπὸς τοῦ συγγράμματος is said at the end of the first lemma by Asclepius and confirmed by Philoponus, who uses a slightly different expression: σκοπὸς τῇ προκειμένῃ συγγράμματι.[316] So this was Ammonius' view. But before the first lemma Philoponus has added a brief introductory passage, in which he gives us the αἴτιον τῆς ἐπιγραφῆς or *explanation of the title*: Εἰσαγωγὴ

Proclus Procleius of Laodicea in Syria at *Suda* Π 2472 (4.210.1-4 Adler), Εἰς τὴν Νικομάχου Εἰσαγωγὴν τὴν ἀριθμητικήν, is lost. So is that by an otherwise unknown Heronas mentioned by Eutocius *in Arch. De sphaer.* 3.120.20-3 Heiberg: ἐν τῷ ὑπομνήματι τῷ εἰς τὴν Ἀριθμητικὴν εἰσαγωγήν. That the anonymous introduction discussed in Ch. XI 3 below is a fragment of this Heronas (Proclus Procleius, who probably is to be dated before the end of the 4th cent. CE, seems too early) is of course entirely speculative.

[311] Asclepius ed. Tarán (1969), sections of Philoponus ed. Haase (1982). The earlier editions of Philoponus' version published by Hoche in the sixties of the last century, said to be unreliable, were not accesible to me.

[312] Tarán (1969) 10, 12-3.

[313] Philoponus in the introduction to his little monograph *De astrolabo* too says that the subject has already been treated by Ammonius, τῷ ἡμῶν διδασκάλῳ (I quote from the repr. of the Greek text in Segonds [1981] 143). Here the isagogical issues are stated quite clearly: the *topic*, viz. the *explanation* (ἐξάπλωσιν—a technical term, see Mansfeld [1994] 149) of the projection of the sphere on the astrolabe etc., what this instrument is *useful* for (χρήσιμος), more *clarity* (πλείονος ... σαφήνειας) than had been provided by Ammonius to make the account more comprehensible for those with no special training in the subject, viz. astronomy (τοῖς μὴ ταῦτα πεπαιδευμένοις—*qualities required of the student*), an ambition comparable to that which impelled Theon to write his *Little Commentary* (above, Ch. X 1 *ad finem*). For Ammonius' astronomical teaching see below, p. 129, complementary note 260.

[314] Pappus says he is a Pythagorean, see above, text to n. 68.

[315] Ascl. 24.1-4 Tarán, Philop. 401.9-10 Haase.

[316] Ascl. 25.63 Tarán, Philop. 405.21 Haase.

ἐπιγέγραπται ὡς πρὸς τὰ γεγραμμένα αὐτῷ Θεολογικὰ ἤτοι Μεγάλα ἀριθμητικά.³¹⁷ This is followed by a pathetic attempt to include a *Vita*-element. Nicomachus is called 'of Gerasa' he tells us, because this is his city of birth. He informs us where Geresa is, and how it came by its name ...

XI 3 *The Anonymous Prolegomena to the* Introductio

This short tract³¹⁸ begins with a *definition* of arithmetic as a theoretical discipline dealing with what is the case with numbers as to their quantities, forms and proportions, as well as to their divisions and combinations. The specific matter it deals with is determinate quantity, consisting of conceptually indivisible minima. We further read of its primary *division into* two *parts*, viz. the theories of plane and of solid numbers, and then of another dichotomous subdivision, viz. into numbers that measure and those that are measured.

Because the *Introductio*, as we have seen, was taught in the Neoplatonist school of Alexandria (and presumably at Athens too), and caught the attention of Boethius, it is safe to assume that the author of these Prolegomena worked in a scholastic Neoplatonist establishment. Also see below, on Pythagorean and Platonic philosophy.

The *theme* (σκοπός) of the present treatise is the treatment of the number that measures, the other kind of number having been treated by Diophantus in the thirteen books of his *Arithmetic*. But the σκοπός of Nicomachus is to instruct us about the number that measures, and in the proem of his book he straightaway speaks by way of a prelude of the *theme* and its *utility* (τὸν σκοπὸν πρότερον καὶ τὸ χρήσιμον προανακρουσαμένος, 2.73.29-74.1 Tannery). Next he

---

³¹⁷ Cf. above, text to n. 58.
³¹⁸ Ed. Tannery (1895), who gave it the apposite title "Anonymi prolegomena ad Introductionem arithmeticam Nicomachi"; the ms. he consulted (*Paris. gr.* 2372) has the heading Περὶ ἀριθμητικῆς. In his "Prolegomena" p. xiii Tannery attributes the piece to a Byzantine scholar perhaps to be dated to the time of Psellus, but I agree with O'Meara (1989) 19 n. 39 that it dates to late antiquity. It is comparable to the *Prolegomena* to the *Suntaxis* for which see above, Ch. X 2. According to D'Ooge *& al.* (1926) 126 it "contains little of interest either to the mathematician or the historian." Tarán (1969) 6 n. 15 agrees: "it contains nothing important either mathematically or philosophically." It will become clear that I believe it to be interesting from the point of view of the history of philosophy and mathematics.

investigates (ζητεῖ) five topics concerning numbers each of which is briefly described, a description which is rounded off with the phrase 'it is Nicomachus' *aim* to teach these subjects in the manner of an introduction' (περὶ τούτων μὲν οὖν σκοπὸς τῷ Νικομάχῳ ὡς ἐν εἰσαγωγῇ παραδοῦναι, 74.26-7). 'In the manner of an introduction'— clearly, the anonymous author wants to insist on Nicomachus' *manner of presentation.*

He continues by advising us that the treatise is also *useful* for our understanding of Pythagorean philosophy (χρησιμεύει δὲ ἡμῖν εἴς τε τὴν Πυθαγορικὴν φιλοσοφίαν, 74.28-9). Things were said to be from numbers by Pythagoras, and a number of arithmological illustrations of this principle are duly provided, mostly concerned with the number seven. This section is again rounded off with a summarizing phrase, viz. διὰ ταῦτα μὲν οὖν τῇ Πυθαγορικῇ φιλοσοφίᾳ *χρήσιμον* τὸ βίβλιον, 75.19-20. But it is also useful for Platonic philosophy , since Plato called the demiurge One (ἕν).[319] It also contributes to the study of nature (φυσιολογίᾳ), Anonymus continues, for many miscarriages occur and many malformed children are born because of the different number of the time concerned.

(The science of) numbers has to be placed before all other mathematical disciplines, because numbers are prior to everything else, as Nicomachus too proves in what follows. Number is incorporeal (proofs provided). Accordingly arithmetic comes first in the *order* of the mathematical disciplines (προτέραν ... τετάχθαι), and canonics (μουσική) comes before astronomy: In the *Great Astronomer*[320] it is shown that the regular motions of the heavenly bodies occur according to rhythm and harmony (76.10-4).

The study of this treatise, viz. the *Introductio,* which is of an introductory nature, has to *come before* (προαναγνῶναι) that of Nicomachus' other *Arithmetic,* to which he gives the *title* (ἐπιγράφει) *Great Arithmetic,* or *Theologoumena.*[321] In this other treatise Nicomachus actually refers to the *Introductio,* thus proving both its *authenticity,* γνήσιον, and the τάξις, i.e. the τάξις τῆς ἀναγνώσεως or *order of study* of the two treatises, as well as their *systematic sequence,* 76.20-4. Finally, the *division into parts* of the work: this is into two books (76.25-6). The contents of each book are then briefly summarized.

---

[319] See below, p. 130, complementary note 319.
[320] See above, text to n. 61.
[321] See above, text to n. 60.

It is clear that this Anonymus knows and applies the scholastic isagogical scheme, inclusive of its technical vocabulary. That his ordering of issues is a bit free is caused, presumably, by his desire to provide an informative summary of the contents of Nico-machus' treatise, whom in fact he follows quite closely. His little tract is a good example of the *ante opus* section of a commentary;[322] though we hear little enough about Nicomachus himself, we are at least given a *catalogue raisonné* of two of his works, and as a sort of bonus even a preview of the section about the number seven (in book II) of the *Theologoumena.*[323]

---

[322] See above, n. 201 and text thereto.
[323] See 60.2-63.5 Pistelli in the abstract at [Iambl.] *Theol. ar.*

CHAPTER TWELVE

CONCLUSION

We may conclude by stating that the evidence available in the various fields and genres of ancient mathematics confirms the development outlined in an earlier enquiry.[324] Ancient mathematics, and especially the teaching of mathematics, did not proceed in splendid isolation, but developed along lines parallelel to the development of general literate culture.

Euclid's works lack introductions, or dedications, and the earliest extant astronomical treatises too begin *in medias res*. This however changed already in the third century BCE. A number of Archimedes' extant works do have letters of dedication which tell us something about their contents in advance. Shortly after 200 BCE the great Hellenistic mathematician Apollonius went much further. We have found that in his great treatise too, just as in early examples of literature in other fields, isagogical issues are used implicitly, that is to say in an unscholastic way, but that he is quite aware of what he is doing. In this context it is most significant that his innuendos could be taken up by the Neoplatonist Ammonius' pupil Eutocius, nine centuries later, and that Pappus too found it worth his while to quote from his general prologue. To pick out only a few further highlights: Heron in the first century CE already wrote introductory works of which the title begins with 'What Comes Before ...', Tὰ πρὸ ... (compare the much later author of the Prolegomena [Tὰ πρὸ ...] to Euclid's *Optica* ascribed to Theon, who felt that an introduction was lacking and had to be supplied). The extant one of these two works of Heron, better known by its Latin title *Definitiones*, is in the first place intended as an introduction to Euclid's *Elements*, though the author also included other material and so broadened the spectrum quite a bit. Ptolemy in the second century CE employs isagogical issues in a sophisticated way, and they are of undeniable importance to him. A century and a half later Pappus in his Commentary on Euclid *Elements* book X uses a number of these issues quite explicitly, and we have

---

[324] Mansfeld (1994); see above, Ch. I.

seen that he also employs them in his *Collectio*. In the latter work, moreover, the existence of corpora of classical astronomical and mathematical writings is attested, as is the way these were taught.

It hardly is a surprise that Proclus' pupil and successor Marinus in his Commentary on Euclid's *Data* is quite scholastic in his use of the isagogical scheme. Finally, we have seen that the full-fledged scholastic scheme is present in several anonymous introductory pieces, almost certainly of Neoplatonist provenance, viz. the Prolegomena to Ptolemy's *Suntaxis* and that to Nicomachus' *Introductio*. Although these late tracts are in themselves of little significance (and a trifle tedious), they are highly interesting because they attest the culmination of a development from the implicitly expressed to the explicitly expressed, and from there to scholastic routine. This development is not different from that in the fields of philosophy, medicine, and so on, and provides additional witness to the fact that by the end of antiquity instruction in mathematics, philosophy and medicine was given by the same people, or at least by people connected with philosophical schools where these various displines were taught.

It is sometimes argued, e.g. by Mme. Hadot, that the mathematicians were philosophers, i.e. that mathematics was no longer an independent discipline already in the early imperial period, if not earlier.[325] This is a view I cannot share. I limit myself to few prominent examples. Take Pappus. The *Suda* indeed calls him a 'philosopher'[326] and so does the author of the late anonymous Commentary[327] at p. 1164.17 Hultsch, but this is an anachronism, notwithstanding Pappus' interest in and knowledge of philosophy (for which see below, Appendix 2). As a matter of fact, at *Coll.* 1.350.28-9 he polemizes in a quite characteristic way against them: 'the philosophers fail to provide proofs and merely affirm something', οὔθ' οἱ φιλόσοφοι δεικνύουσιν, ἀλλ' ἀποφαίνονται μόνον.[328] This is not the way of speaking of a person who considers himself a philosopher. Furthermore, at *Coll.* 3.1022.5-6 he distinguishes the

---

[325] E.g. Hadot (1984) 252-61, who provides a fast survey of mathematical literature from Geminus to late antiquity.; also see Decorps-Foulquier (1992) 54, 56-8 on 'the philosopher Serenus' in a fragment found in certain mss. of Theon of Smyrna, Heiberg (1893) pp. xviii-xix (on Serenus see above, nn. 8, 25, 142). For Ptolemy see above, n. 226.
[326] See below, n. 356; cf. Hadot (1984) 257.
[327] Cf. above, Ch. X 2.
[328] See below, n. 355 and text thereto.

philosophers from the mathematicians. Also compare Heron of Alexandria's scathing comment on the disagreement among the philosophers at the beginning of the *Belopoiica*.[329] Heron, too, clearly is not a member of the philosophical profession. On the other hand the 'philosopher Hierios' cited by Pappus at *Coll.* 1.24.3 obviously was someone who practised mathematics in a professional way. So it is plain that some philosophers practised and taught mathematics, while on the other hand persons can be recognized who were mathematicians, not philosophers, though they were to some extent at home in the world of philosophy. They were civilized people who had received a good education. For late antiquity Mme. Hadot's view is of course entirely correct.

The alchemical oath attributed to 'Pappus, philosopher' (Παπποῦ φιλοσόφου ⟨ὅρκος⟩),[330] even if genuine, does not prove he was a philosopher either, and does so for the same reason. Authenticity is admitted as a possibility by Bulmer-Thomas, and Mme. Hadot emphatically argues in its favour.[331] But I find the 'cherubic chariots' and 'angelic throngs' (ἁρμάτων χερουβικῶν and ταγμάτων ἀγγελικῶν) carrying and accompanying the Creator to whom the oath is sworn hard to stomach.[332] It could be argued that the sentence at the end containing these Christian ingredients was added later (especially the cherubim are remarkable, for angels—though hardly throngs of them—can be paralleled from pagan literature). Even so, I believe that it is far more plausible that we are dealing with a not so pious fraud. One only has to recall the pseudigrapha attributed to Democritus, or Theophrastus, or Archelaus, etc., in the alchemical literature, even in the manuscript containing the oath ascribed to Pappus.

We may finish by stating that the mathematical evidence investigated in the present enquiry increases our knowledge in several ways. Abundant parallels are found for ways of presentation and methods of teaching known from various other fields,

---

[329] Above, n. 166 and text thereto.

[330] Berthelot and Ruelle (1888) 2.27.18-28.4 (transl. 3.29-30). According to their report only found in *Marc. gr.* 299, dated by them to the 11th cent. (*ibid.* 2.2).

[331] Hadot (1984) 257, Bulmer-Thomas (1974) 301.

[332] The few parallels for these specific formulas I have found are all in Christian authors, and I have failed to find a single one for their occurring together. What is more, I have found only one further instance of the 'cherubic chariots', viz. John of Damascus, *Homilia in ficum arefactam*, Migne *PG* 96.576.31, ὁ ἐπὶ Χερουβικῶν ἁρμάτων ἐποχούμενος, a formula pertaining to Christ.

and some among these parallels are quite early. Perhaps the most spectacular from a chronological point of view are Apollonius' proems to the *Conica* as a whole and to the individual books, in which isagogical issues play such a remarkable role. These are much earlier than the early material taken into consideration in another book by the present writer.[333] Moreover, the evidence provided by Apollonius is far richer than the precedents to be found in still earlier authors such as Aristotle.[334]

---

[333] Above, n. 1.
[334] See above, n. 10 and below, pp. 122-3, complementary note 11.

APPENDIX 1

THE TITLE OF PTOLEMY'S ASTROLOGICAL TREATISE

For the book-title Ἀποτελεσματικά Boll and Boer follow the titles of the individual books in the best ms. (*Vat. gr.* 1038, 13th cent.)[335] The anonymous Commentary discussed above, Ch. X 2, provides yet another variation, viz. ἐν τοῖς πρὸς Σύρον γενεθλιακοῖς τέτρασι βιβλίοις (γενεθλιακοῖς is not entirely correct, since individuals are only dealt with in books III-IV). Lyd. *Mens.* 155.4-6 Wuensch refers to *Apotel.* 92.7 Boll and Boer in the words ὁ δὲ Πτολεμαῖος ἐν τοῖς πρὸς Σύρον αὐτῷ γραφεῖσι προστίθησι κτλ., as if no other works had been dedicated to this person. Nicephorus Gregoras (13th-14th cent.), *Hist. byz.* 25.11, p. 3.32.16-7 Bekker, speaks of τὴν Πτολεμαίου ἀποτελεσματικὴν τετράβιβλον. Other varieties found in the extant mss., among which Τετράβιβλος, are cited in the *app. crit.* of Boll and Boer (1940) 1. The *Fihrist*, Dodge (1970) 2.640, also calls it 'the Four'. For the title Ἀποτελεσματικά ascribed to Manetho in the *Suda* see above, n. 61; cf. *Suda s.vv.* Helikonius E 852 (2.247.8 Adler), Zoroaster Z 159 (2.514.18), Paulus Alexandrinus Π 810 (4.69.19-20). The compilation of Hephaestion of Thebes (published ca. 315 CE), books I-II of which contain numerous extracts from Ptolemy's treatise, is published with the title *Apotelesmatica* by Pingree (1973). I note in passing that it begins with the words Σὺν θεῷ ἡμῖν σκοπὸς ἐνθάδε—an early instance of this terminus technicus right at the start of a treatise.

Erotianus *Voc. hipp.* 5.4 Nachmanson lists a ἑξαβίβλος πραγμα-τεία by Philinus. Galen, *Diff. febr.* 7.311.3-4 Kühn mentions a τετράβιβλον [scil., πραγματείαν] περὶ τῶν ἐν τοῖς σφυγμοῖς αἰτίων (so Περὶ τῶν κτλ. is the real title), and *Meth. med.* 10.37.18 sarcastically speaks of ἑκατοντάβιβλοι πραγματεῖαι; here we are dealing with adjectives not substantives. But Paul of Aegina (7th cent.) procœm., 1.4.6 Heiberg refers to ἡ ... Ἑβδομηκοντάβιβλος αὐτοῦ τοῦ Ὀριβασίου, and the *Suda* lemma on Hippocrates (I 564, 2.663.3 Adler) mentions Hippocrates' πολυθρύλλητος καὶ πολυθαύμαστος

---

[335] Also see Boer at Ziegler *& al.* (1959) 1831-8.

Ἑξηκοντάβιβλος. Phot. *Bibl.* cod. 127, 95b5-7 Bekker refers to Eusebius' *Vita Constantini* as ἡ εἰς Κωνσταντῖνον τὸν μέγαν βασιλέα ἐγκωμιαστικὴ τετράβιβλος, and *ibid.* lines 16-7 refers back to it in the words ἐν ταύτῃ αὐτοῦ τῇ τετραβίβλῳ. Similar Photian examples: cod. 85, 65b, ἡ εἰκοσάβιβλος αὕτη ἡ Κατὰ τῶν Μανιχαίων πρὸς Ἀχίλλιον, cod. 140, 98a, τοῦ αὐτοῦ ἁγίου ἡ Κατὰ Ἀρείου καὶ τῶν αὐτοῦ δογμάτων πεντάβιβλος. We note that in these references to a 'manybook' further information concerning the contents or title is often included. Stegemann (1939) 6-7, followed by Gundel and Gundel (1966) 206, defends the title Τετράβιβλος (though with some hesitation) with the odd argument that Ptolemy wanted to distinguish his treatise from the astrological poem by Dorotheus of Sidon (on him, 1st half of 1st cent. CE, see Stegemann *ibid.* 1-5, Gundel and Gundel *ibid.* 117-20, Pingree [1978]). Numerous fragments in Greek or translated into Latin are extant; so is an (interpolated) Arabic translation to be dated to ca. 800 (itself translated from the Pahlavi), see Pingree (1976) who provides the *editio princeps* of the Arabic text and an English version, and adds the fragments. For these Pahlavi and Arabic translations in their habitat see Pingree (1997). This work does have five books, and is indeed called 'the Book of Five' in the *Fihrist*, Dodge (1970) 641; an-Nadim subsequently lists a sixth, seventh and even sixteenth 'section', but this will be a mistake. Though Dorotheus' work at some time acquired the designation 'Fivebooks', this would be utterly strange as the original title of a poetical work. So much is admitted by Stegemann (1939) 6, who however defends the title found in the *Fihrist* though he knows that Firmicus Maternus, *Math.* 78.3-5 Kroll and Skutsch speaks of Dorotheus' *Apotelesmatica verissimis et disertissimis versibus.* Pingree in his edition simply calls it *Carmen astrologicum,* and does not give it a title in his (1978) encyclopedia article.

As Carlos Steel points out to me, Willem van Moerbeke translated the title as *Iudicalia ad Syrum* (see Vanhamel [1989] 369]), which as it would seem supports Ἀποτελεσματικὰ πρὸς Σύρον, not Τετράβιβλος.

Ptolemy's Ἀποτελεματικὰ πρὸς Σύρον δ'—as I suppose the proper title will have looked like—apparently became sufficiently famous to be called by the designation Τετράβιβλος alone. Cf. ἡ Πεντά-τευχος (earliest occurrence in Ptolemaeus the Gnostic's *Ep. ad Floram* 4.1, 2nd cent. CE, which has escaped Bogaert [1997], a paper

which is otherwise a useful overview of part of the evidence for
-τευχος/*ticus*) = our Pentateuch, Ὀκτάτευχος for the first eight books
of the Old Testament but also as the title of a book ascribed to
Ostanes (Philo of Byblus *ap.* Eus. *PE* 1.10.53, text printed at
[Democr.] Fr. 300.13a DK), or even our 'Bible'. For the remote
possibility that the *Suda* referred to the work as τὰ δ΄ βιβλία [=
Τετράβιβλος] Πτολεμαίου see above, n. 249.

My hypothesis is that Dorotheus' epic came to be called
Fivebooks by the Arabs (or was even so entitled in their Greek mss.
already) to distinguish it from the *Fourbooks*. Possibly the *Apotelesmatica* in four books had come to be called *Tetrabiblos* to distinguish
it from the *Megalè Suntaxis* in thirteen books.

PAPPUS AND THE HISTORY OF PLATONISM

In this section I want to discuss three passages in the *Sunagôgè* which are of interest for the history of Platonism (and Platonizing Pythagoreanism). As far as I know they have been overlooked by historians of philosophy, while naturally they have proved to be of little interest to historians of mathematics.[336] Even so, I believe that they are important for the light they shed on the history of Platonism in the imperial period. Treatment of a number of equally interesting passages in the Commentary on Euclid *Elements* X must regretfully be postponed till another occasion.

A not entirely unjustified view which still is quite wide-spread (though less wide-spread than it used to be) is that there is a major trend in Middle Platonism, chiefly represented by Alcinous and Numenius, which helped to prepare the way for the complicated Neoplatonist system of Plotinus and the even more complicated ones of the Late Neoplatonists.[337] The formula 'Middle Platonism' presupposes the existence of something to be designated Neoplatonism, and is as questionable as, say, 'Middle Comedy'. 'Neoplatonism' is of course a neologism itself, involving a evaluative

---

[336] Knorr (1986) 357 on Pappus' references to Plato in the *Sunagôgè* (of which he notes only one, viz. that about the harmonic mean in the *Timaeus*, for which see below) and on the Commentary on *Elements* X is insufficient, and his suggestion that Pappus got the information to be found in the *Sunagôgè* via "commentators like Geminus and others, conversant with a syncretistic form of Platonism" and hence that "Pappus himself might not be fully aware of the ultimate provenance of his views" is not good enough, as we shall see. Knorr moreover has missed Pappus' reference to Nicomachus.

[337] As appears for example from the title *The Handbook of Platonism* given by Dillon (1993) to his translation of Alcinous' *Didascalicus*. One may also think of Willy Theiler's celebrated formula *Vorbereitung des Neuplatonismus,* or of tendencies in the account of Merlan (1967). Still, Dillon (1977) xiii and elsewhere argues that matters are less simple. One only has to think of the controversies concerning Calcidius *In Timaeum,* for which see Dillon *ibid.* 401-8. Donini (1982) 11-27, in his splendid evaluation of the history of the scholarship concerned with the philosophies of the 1st cent. BCE and the 1st-3rd cent. CE, insists that the teleological approach is misleading, and *ibid.* 100-59 demonstrates how complex a phenomenon 'il platonismo medio' really is. Also see Manfeld (1982).

judgement, like 'Middle Platonism'. But we are stuck with this terminology, and I shall use it myself.

To be sure, it is generally admitted that there were also other currents in so-called Middle Platonism, which however are, or were, considered to have been less successful. In a sense they certainly were, but the value judgement involved is very much a question of insight by hindsight. The development which so to speak in a teleological way paved the way for the advent of Neoplatonism is a modern construct, which is heavily indebted to the geneticist, or developmental, paradigm.[338] But cultural development should not be conceived in terms of the development of the embryo. I do not deny that Plotinus was indebted to his Platonist (and Neopythagorean) predecessors, but believe that the arrow points the other way, that is to say backwards. What is important is Plotinus' reception of what, with some hesitation, we may call the 'traditions' concerned with the interpretation of Plato, and this to a quite impressive degree amounts to selection as well as creative interpretation. *Quidquid recipitur ad modum recipientis recipitur.*

My enquiry will be restricted to the reception of the *Timaeus.* It is well known that Plato in this dialogue argues that the cosmos is fabricated by a supreme God, most of the time called 'the Demiurge'—but designated 'builder' (τεκταινόμενος) at *Ti.* 28c5, 'Intellect' at 39e7, and 'the Maker and Father of this universe' at 28c3—, who imposes forms and structure on the unwilling Receptacle by looking at the transcendental Form of Living Being which contains the Forms of the other Living Beings. It is also well known that later Platonists regarded the Platonic Ideas, or Forms, as objects of the Divine Intellect on the same ontological level, or even placed them as its 'thoughts' in the, or a, Divine Intellect itself, as Plotinus too was to do (Porphyry at first disagreed with Plotinus, but was won over in the end, *VP* 18). Alcinous and other Middle Platonists such as Numenius multiplied the number of Gods, or Intellects. Alcinous' First God/Intellect contains the Ideas; his only activity, if that is what it may be called, is to awaken the Second God. The demiurgic task of making the universe is taken over by this Second God/Intellect, inspired and prompted by the First.[339] In the

---

[338]   For this paradigm see Crombie (1994) 3.1547 ff., and Mansfeld (1998b).
[339]   Alcin. *Did.* ch. 10, see Dillon (1977) 282-3, Donini (1982) 106-7. But traces of the less sophisticated view remain, see *Did.* 163.13 and 172.5 Hermann.

case of Numenius (in his treatise *On the Good*) the First God/ Intellect is 'inactive' (ἀργός, *ap.* Eus. *PE* 11.18.8), while the Second God/Intellect generates the Third, the Demiurge who constructs the universe, and does so by so to speak dividing itself into two.[340] These complicated approaches to the relation between the intelligible world and the material world of sense-perception are further refined by Plotinus. All of this took place quite a long time before Pappus, who as we know has to be dated to the first half of the 4th century. Yet, when one reads the *Sunagôgè*, it looks as if no such thing had happened. If we had only Pappus and, say, Proclus minus his historical overviews, our impression of the history of Platonism would be quite different.

This is clear from three passages in the *Sunagôgè*. In the first of these (5.19), the introduction to the second part of book V which deals with the regular convex solids, he writes as follows (my italics):[341]

> *The* philosophers say that it is plausible that *the First God and Demiurge of all things*, choosing the most beautiful of all shapes, gave the cosmos the shape[342] of a sphere. They describe the natural characteristics[343] of the sphere, and add that the sphere is the

---

[340] For Numenius see esp. Frs. 11-13 and 16-17 Des Places. There is a difficulty here, since the Second Intellect and the Demiurge are said by him to be 'one' in some sense of the word 'one', Fr. 11 Des Places *ap.* Eus. *PE* 11.18.3, ὁ θεὸς μέντοι ὁ δεύτερος καὶ τρίτος ἐστὶν εἷς, which explains why the fragments for the most part speak of two Gods only; see Donini (1982) 142 and Frede (1987b) 1057-70, whose explanation I have followed in the text. This Second-and-Third God is the result of an original exegesis of the δεύτερον ... πέρι τὰ δεύτερα and τρίτον πέρι τὰ τρίτα of [Plato] *Ep.* 2.312e, see Donini, *loc. cit.* Also cf. below, n. 362 and text thereto. O'Brien (1992) 333 points out that Numenius' doctrine (*sine nomine auctoris*) of the 'idle God' is criticized Plot. *Enn.* 2.9.1.27-9. Atticus rejected a multiplication of Gods of this sort, see Fr. 4 Des Places *ap.* Eus. *PE* 15.6.2-17, and e.g. Donini (1983) 115.

[341] 1.350.20-30 Hultsch, τὸν πρῶτον καὶ δημιουργὸν τῶν πάντων θεόν οἱ φιλόσοφοί φασιν εἰκοτῶς τῷ κόσμῳ σχῆμα περιθεῖναι σφαιρικὸν ἐκλεξάμενον τῶν ὄντων τὸ κάλλιστον, τά τε πρόσοντα τῇ σφαίρᾳ φυσικὰ συμπτώματα λέγοντες ἔτι καὶ τοῦτο προστιθέασιν ὅτι πάντων τῶν στερεῶν σχημάτων τῶν ἴσην ἐχοντῶν τὴν ἐπιφάνειαν μεγίστη ἐστὶν ἡ σφαῖρα. τἄλλα μὲν οὖν ὅσα προσεῖναι λέγουσιν αὐτῇ πρόδηλά τε ἐστι καὶ παραμυθίας ἐλάσσονος δεῖται, τὸ δ' ὅτι μείζων ἐστὶ τῶν ἄλλων σχημάτων οὔθ' οἱ φιλόσοφοι δεικνύουσιν, ἀλλ' ἀποφαίνονται μόνον, οὔτε παραμυθήσασθαι ῥάδιον ἄνευ θεωρίας πλείονος. For the formula οἱ φιλόσοφοί φασιν see below, n. 355.

[342] σχῆμα περιθεῖναι is standard later Greek, see e.g. Gal. *UP* 3.471.2 Kühn and *PHP* 9.8.8.~ p. 157 unr column

[343] The formula φυσικὰ συμπτώματα is rare. Its earliest occurrence is Arist. *GA* 4.10.777b9, on why certain animals enjoy long life; this is explained e.g. *Long.* 4-5.466a15 ff.: the living being is 'by nature humid and warm'. For the meaning 'symptom' (such as coughing in certain diseases) e.g. Gal. *Loc. aff.*

greatest[344] of all the figures which have the same surface (as the sphere). The other characteristics they ascribe to the sphere are clear enough and need little or no explanation. However, that the sphere is greater than the other figures[345] *is not proved by the philosophers but merely affirmed by them.* It is not so easy to explain this without appealing to a theoretical enquiry which goes a great deal further.

Pappus continues by reminding us that in the preceding chapters of book V he has proved (according to the *mos geometricus,* or γεωμετρικὸς τρόπος, of course) that the circle is the greatest of all regular planes with their vertices on the same circumference, and states that in what follows he will do the same for the sphere and the regular convex solids of which the sphere is the including limit. But *all* these regular solids will have to be treated:[346]

> These are not only the five shapes found in the most divine Plato,[347] that is to say the tetraeder and hexaeder, octaeder and dodecaeder, and the icosaeder as fifth,[348] but also those

---

8.325.15 Kühn. Iambl. *CMSc.* 75.13-5 Festa argues that the Pythagoreans were less interested in difficult mathematical theorems than in those providing an insight in the order (of nature), or in τι σύμπτωμα φυσικόν. As to the 'natural chararcteristics' (for which see also below), already Parmenides' Being (28B8.42-3 DK) is 'perfect' and resembles a 'well-rounded sphere', τετελεσμένον... πάντοθεν, εὐκύκλου σφαίρης ἐναλίγκιον ὄγκῳ. Plato's spherical cosmos possesses 'the most perfect of all figures', πάντων τελεώτατον ... σχημάτων (*Ti.* 33b). Perfection of the circle (and of circular motion) often in Aristotle, e.g. *Cael.* 269a20, 286b22-3, *Phys.* 264b27-8. Alexander Polyhistor quoted D. L. 8.35 (= Anon. Pyth. 58C3 DK, 1.463.24-5) said he had read in the *Pythagorean Hypomnemata* that the sphere is the most beautiful solid and the circle the most beautiful plane figure: the *topos* is attributed to the (early) Pythagoreans.

[344] I.e. has the greatest volume.

[345] Note that these figures can be inscribed in it, as Euclid proceeds to do in *Elem.* XIII, constructing the sphere by rotating a half circle.

[346] 1.352.11-5, [...] ταῦτα [*scil.* πολύεδρα] δ' ἐστὶν οὐ μόνον τὰ παρὰ τῷ θειοτάτῳ Πλάτωνι πέντε σχήματα, τούτεστιν τετράεδρόν τε καὶ ἑξάεδρον, ὀκτάεδρόν τε καὶ δωδεκάεδρον, πέμπτον δ' εἰκοσάεδρον, ἀλλὰ καὶ τὰ ὑπὸ Ἀρχιμήδους εὑρεθέντα τριακαίδεκα τὸν ἀριθμὸν κτλ.

[347] Of this celebratory formula, which occurs twice in the *Sunagōgè,* I have found thirteen other examples, mostly in Neoplatonist authors, but it occurs already at Gal. *UP* 4.266.4-5 Kühn and *PHP* 9.9.3, and Athen. *Deipn.* 10.55.

[348] The tetraeder, hexaeder, octaeder and icosaeder are the ultimate constituents of the four physical elements (fire, earth, water, air) in the *Timaeus,* while the dodecaeder so to speak may be inflated to the shape of a ball (cf. *Phd.* 110b and e.g. Iambl. *VP* 247) and is the figure for the cosmos as a whole at *Ti.* 55c. Correctly formulated by Gal. *Comp. Tim.* 10a Kraus and Walzer: 'ignis species figura ignea [mistranslation of πυραμίς] est, et terrae species figura cubica, et aquae species ea figura est quae viginti bases habet, et aeris species ea figura est quae octo bases habet. Deinde dixit: Etiam alia forma exstat propter totum mundum exstructa; iudicavit autem figuram quae duodecin

discovered by Archimedes, which are no less than thirteen in number.[349]

It is of some interest to observe that Euclid believed he had proved that there can be *no more than* five regular convex solids.[350] Archimedes' discovery of the semi-regular convex solids therefore created a problem for Platonists who would believe that Plato and Euclid had said the final word on the subject. That this may have been the case is suggested by a passage in Heron.[351] Its formulation is confusing (possibly because of an accident in the transmission), since it wrongly states that Archimedes added eight solids to Plato's five, the only ones accepted by Euclid. But Heron at any rate refers to a view according to which Plato already 'knew' two of Archimedes' solids, viz. two tetradecaeders (of the latter's three).[352] The statement that Plato 'knew' presumably goes back to a comment, or Commentary, on the *Timaeus* which attempted to find Archimedes' discovery in Plato (a quite normal exegetic ploy).[353] Now if Plato 'knew' two of Archimedes' semi-regular solids, he knew the principle according to which they are to be constructed, so potentially 'knew' all of them. There is some further evidence for references to Euclid in the commentary tradition. Gal. *Comp. Tim.* 3a Kraus and Walzer, speaking of the two *mesotètes* of solids

---

bases habet'. Similar but longer version Alcin. *Did.* 12, 168.8-24 Hermann. Useful n. 241 at Whittaker (1990) 29, who points out that of the five technical terms Plato only uses pyramid, and that the others appear in a Platonic context for the first time in Timaeus Locrus and Plutarch.

[349] For these Archimedean semi-regular solids Pappus is our main source. The texts of Pappus, of the scholia on this passage of Pappus, and of Heron (for the latter see below, n. 352 and text thereto) concerning Archimedes' polyedra are also printed at Mugler (1972) 202-7.

[350] *Elem.* XIII demonstr. 18, epimetrum 113 ff., 135 ff. (referred to by Heron, see below n. 362). Note the fourth problem of Pappus *Coll.* book III at 1.132.1-2 Hultsch (my italics): 'to inscribe *the* five polyedra in a given sphere'; in the sequel Archimedes' solids are not mentioned.

[351] To be dated, as we recall, to the 1st cent. CE.

[352] Heron *Def.* 104, Εὐκλείδης μὲν οὖν ἐν τῷ ιγ' τῶν Στοιχείων ἀπέδειξε, πῶς τῇ σφαίρᾳ τὰ πέντε ταῦτα σχήματα περιλαμβάνει· μόνα γὰρ τὰ Πλάτωνος οἴεται. Ἀρχιμήδης δὲ τριακαίδεκα ὅλα φησὶν εὑρίσκεσθαι σχήματα δυνάμενα ἐγγραφῆναι τῇ σφαίρᾳ προστιθεὶς ὀκτὼ μετὰ τὰ εἰρημένα πέντε· ὧν εἰδέναι καὶ Πλάτωνα τὸ τεσσαρεσκαιδεκάεδρον, εἶναί τε τοῦτο διπλοῦν, τὸ μὲν ὀκτὼ τριγώνων καὶ τετραγώνων ἐξ σύνθετον, ἐκ γῆς καὶ ἀέρος, ὅπερ καὶ τῶν ἀρχαίων τινὲς ἤδεσαν, τὸ δὲ ἕτερον πάλιν ἐκ τετραγώνων μὲν ὀκτώ, τριγώνων δὲ ⟨2⟩, ὃ καὶ χαλεπώτερον εἶναι δοκεῖ.

[353] It is far less likely that Archimedes said so himself. As to the ploy one may for instance think of the efforts to find Aristotle's categories and syllogistic in Plato, see e.g. Alcin. *Did.* ch. 6.

and the single *mesotès* of planes in the *Timaeus* (for more on these means see below), adds: 'Quod iam Euclides exposuit'.

However we should return to the Pappus passage, a sleeping beauty which I shall attempt to kiss. In the first place, someone who says '*the* philosophers fail to provide proofs and merely affirm something'[354] evidently does not consider himself to be a philosopher.[355] This is of some importance because in later sources Pappus is called 'the philosopher', clearly an anachronism.[356] In the second place he demonstrates his familiarity with a prominent philosophical doctrine. He evidently admires Plato, whom he calls 'the most divine' among the philosophers, and is aware of the fundamental part played by the five regular convex solids in the cosmology of the *Timaeus*. In the third place, an even more interesting fact (at least from my point of view) is that he says that *the* philosophers affirm that the First God is the Demiurge of all things.

This is correct with regard to the *Timaeus*, but entirely incorrect with regard to those Middle Platonists who introduce two Gods, of whom the First merely inspires the Second who then functions as Demiurge. It is of course also false with regard to Plotinus and whoever followed him. But it is strikingly correct with regard to a

---

[354] Parallels for this contrast between affirming and proving e.g. Plu. *Plat. quaest.* 1006C, S.E. *M.* 8.15, Orig. *C. Cels.* 3.73, Ioann. Chrysost. *De paenit.* Migne *PG* 49.34011-3, Simpl. *in Cael.* 678.21-2 Heiberg.

[355] The formula οἱ φιλόσοφοί φασιν at the beginning does not yet imply this. See Epict. *Diss.* 4.1.173-4, referring to philosophical views he shares: παράδοξα μὲν ἴσως φασὶν οἱ φιλόσοφοι, καθάπερ καὶ ὁ Κλεάνθης ἔλεγεν, οὐ μὴν παράλογα, Clem. *Strom.* 7.5.28, ὡς αὐτοί φασιν οἱ φιλόσοφοι, Porph. *Ad Marc.* 28.8, διό φασιν οἱ φιλόσοφοι οὐδὲν οὕτως ἀναγκαῖον κτλ., Athan. *Inc. verb.* 41.5, τὸν κόσμον σῶμα μέγα φασὶν εἶναι οἱ τῶν Ἑλλήνων φιλόσοφοι, καὶ ἀληθεύουσι λέγοντες, Philop. *in Cat.* 65.10 Busse, τὴν πρώτην ὕλην φασὶν οἱ φιλόσοφοι ἀσώματον εἶναι τῷ οἰκείῳ λόγῳ κτλ. Compare the equivalent formula οἱ φιλόσοφοι λέγουσι. Plutarch for instance may use it to indicate philosophers he disagrees from, without implying that he prefers not to be called a philosopher himself (e.g. *Coni. praec.* 142E, *Garr.* 504B). It is several times found in Epictetus, e.g. *Diss.* 1.25.32 (objection of a dialectical opponent), 2.1.25 (Stoic doctrine cited with approval, cf. 2.14.11). Gal. *Dieb. decret.* 9.754.11-2 Kühn uses it of philosophers one may disagree with. Plot. *Enn.* 2.9.1.4 says that the doctors would express themselves correctly if they were to speak as the philosophers do (ἔλεγον ἂν ὀρθῶς, καθάπερ οἱ φιλόσοφοι λέγουσι). Philop. *in An.* 588.10-3 Wallies likewise contrasts physicians with philosophers, and so does Olymp. *in Cat.* 138.14-8 Busse.

[356] *Suda s.v.* Theon, Θ 205, 2.702.11 Adler, Πάππῳ τῷ φιλοσόφῳ, and *s.v.* Pappos, Π 265, 4.26.3 Adler, Πάππος, Ἀλεξανδρεύς, φιλόσοφος; see further above, Ch. XII.

fellow-student of Plotinus, Origen the Platonist, who wrote a treatise entitled *Only the King is Maker*, Ὅτι μόνος ποιητὴς ὁ βασιλεύς.

'King' as designation of the highest principle is derived from [Plato] *Ep.* 2.312e. I believe that the ποιητής of Origen's title is equivalent to 'Demiurge', and that he has the well-known phrase at the beginning of the main part of the *Timaeus* in mind, viz. 'The *Maker* and Father of this universe it is a hard thing to find, and having found him it would be impossible to explain him to everyone' (*Ti.* 28c; famous formula, often discussed, and quoted as a purple passage Stob. *Ecl. eth.* 2.1.15).[357] 'The Maker and Father of this universe' can only apply to Plato's one and only Demiurge; the hoary designation 'Father' emphasizes that the 'Maker' is the Supreme God (for the verb ποιεῖν in this context see *Ti.* 31b, 34b, 35b, 37d, 38b, 38c). This indeed is how Plutarch read the phrase. But he wondered whether 'Father' and 'Maker' (note the inverted order) pertain to different aspects of the Demiurge's activity, asking himself (my italics) 'why did he call *the Supreme God* Father and Maker of all things?' (*Plat. quaest.* 1000E, τί δήποτε τὸν ἀνωτάτω θεὸν πατέρα τῶν πάντων καὶ ποιητὴν προσεῖπεν;)[358] Atticus Fr. 4 Des Places *ap.* Eus. *PE* 15.6.2-17 uses the formula 'Father of all things' (ὁ πατὴρ ... τῶν πάντων, 6.4) for what Plato, introducing the speech of the Demiurge to the younger gods, calls 'he who produced this universe' (*Ti.* 41b), and speaks of 'the power of the Maker of the universe' (τοῦ παντὸς ποιητῇ δύναμιν, 6.7). What is more, he calls

---

[357] The important phrase at *Ti.* 28c, τὸν μὲν οὖν ποιητὴν καὶ πατέρα τοῦδε τοῦ παντὸς εὑρεῖν τε ἔργον καὶ εὑρόντα εἰς πάντας ἀδύνατον λέγειν, is cited in Cornford's translation, slightly modified. Also compare *Ti.* 37c, ὁ γεννήσας πατήρ, and the beginning of the Demiurge's speech, *Ti.* 41a: 'the works of which I am Demiurge and Father, having come into being through me, are indestructible as long as I am unwilling (*scil.*, to destroy them)', ὧν ἐγὼ δημιουργὸς πατήρ τε ἔργων δι' ἐμοῦ γενόμενα ἄλυτα ἐμοῦ γε μὴ ἐθέλοντος. Note that I have junked the comma after ἔργων. Also cf. *Ti.* 42e, 71d (the "Father" of all things is also that of the younger gods). See further below, p. 131, complementary note 357.

[358] The second of his *Platonic questions* is devoted to this issue. For the meaning of the formula in Plutarch see also *ibid.* 1001B, εἰκότως ἅμα πατήρ τε τοῦ κόσμου, ζῴου γεγονότος, καὶ ποιητὴς ἐπονομάζεται, and *De fac.* 927A. See further the excellent exposition of Runia (1986) 107-11, who lists the epithets the Demiurge receives in the *Timaeus*, counts no less than 41 instances of the formula 'Maker and Father' (or its converse) in Philo of Alexandria, and shows that Philo was aware of its Platonic provenance. Also compare Ferrari (1995) 261: "Plutarco, molto piu fedele di Numenio alla lettera del testo platonico, non sembra avere dubbi sul fatto che il dio supremo è contemporaneamente anche il dio demiurgico."

him 'the greatest King' (παμβασιλεύς, *ibid.*, 6.12). Apuleius' view is quite similar to that of Atticus. He calls God 'unus'[359] and 'genitor rerumque omnium exstructor' (*De Plat.* 191); the latter formula obviously translates Plato's ποιητὴν καὶ πατέρα τοῦδε τοῦ παντός, and in fact the rest of Plato's sentence (*Ti.* 29c), about the God who is hard to find and difficult to explain to all, is not only translated in the sequel but even quoted in the original Greek (*ibid.* 191). Also see *De Plat.* 204, on the first of the three kinds of Gods (my italics): 'unus et solus summus ille, ... quem *patrem et architectum* huius divini ordinis supra ostendimus'. Quoting the all-important phrase at [Plato] *Ep.* 2.312e in the original Greek he also calls this God by the name of βασιλεύς, *Apol.* 64.5.[360]

I note in passing that a scholion to book XIV of Epic. *On Nature* [29] [26][361]—possibly deriving from a passage in Epicurus himself —calls the Platonic Demiurge ὁ συνθέτης (a rare term, better known as meaning 'one who puts words together', 'prose-writer'), viz. the 'putter together' of the Platonic figures and solids criticized by Epicurus. This may be justified by the appellation ὁ συνθείς for the Demiurge at *Ti.* 33d2. Still, the sarcastic exploitation of the ambiguities involved in the Greek words is excellent: a 'prosaic' assembler instead of a 'poetic' Demiurge. But we should return to the Platonists.

Alcinous *Did.* ch. 10, 164.40-65.4 Hermann reserves the designation 'Father' for the First God, but does not call the Second God, who 'imposes order on all of nature in this world', by the name of Maker, though it is clear that he plays the rôle of Plato's Demiurge. Numenius' interpretation of the formula τὸν μὲν οὖν ποιητὴν καὶ πατέρα τοῦδε τοῦ παντὸς according to Proclus in his extensive exegesis of Plato's formula (*in Tim.* 1.299.10 ff. Diehl) involves a distinction between the Platonic Father (called 'Grand-father', πάππος, by Numenius) and the Platonic Maker (called 'Son' or 'Descendant', ἔκγονος), the universe being the 'Grandson' or rather 'Descendant' (ἀπόγονος, *in Tim.* 1.303.28-9).[362] Whatever the

[359] See Beaujeu (1973) 256: 'le dieu par excellence'.
[360] See Beaujeu (1973) 256-7, 271, and for more details Hijmans (1987) 422-4, 436-9.
[361] Arrighetti (1973) 270, *in appar.* For the text (*PHerc.* 1148 col. xxxviii Leone) see Leone (1984) 62, for the interpretation *ibid.* 69-7 with n. 672.
[362] Procl. *in Tim.* 1.303.27-304.7 = Num. Fr. 22 Des Places. See Frede (1987b) 1061, who *ibid.* 1069 argues that Numenius may have said this somewhere else, i.e. not in the treatise *On the Good* from which the extensive fragments

correct interpretation of this obscure and to some extent mythologizing terminology (Kronos—Zeus?) and supposing, of course, that Proclus is right in seeing Numenius' phrase as an exegesis of Plato's formula, it seems to follow that, unlike Plutarch, Numenius distinguished individuals not aspects of the same individual.

One should recal that the 'inactive First God' of Num. Fr. 12 is called 'King' by him (τὸν μὲν πρῶτον θεὸν ἀργὸν εἶναι ἔργων ξυμπάντων καὶ βασιλέα, ap. Eus. PE 11.18.8).³⁶³ The simplest explanation of the meaning of Origen's title in my view is that it expresses disagreement with Numenius' novel interpretation of Plato's phrase, which naturally entails that he rejected his Two or Three Gods distinction.³⁶⁴

---

cited above, n. 340, derive. For Numenius' term ἔκγονος see Schol. vet. in Iliad. 5.813, ἔκγονος ὁ υἱός, and Schol. in Soph. Aiacem 842a, ἔκγονος καὶ ἔγγονος διαφέρει. ἔκγονος ὁ υἱός. But note that both ἔκγονος and ἀπόγονος may be used more loosely: more or less remote 'descendant', see LSJ s.vv. Perhaps this allows us to interpolate an entity between the 'Grandfather' and the 'Son', viz. the Second God as father of the 'Son' and grandfather of the 'Grandson'; the 'Grandfather' cited by Proclus then would be the grandfather of the 'Son'. Alternatively, we may perhaps interpolate an entity between the 'Son' and the 'Grandson'.

³⁶³ On the hierarchy of 'Kings' and the low position of the Demiurge of the cosmos in late Neoplatonism see Hadot (1978) 112-4. I still have not entirely come round to her well-argued view that the Demiurge of Hierocles the Platonist (ca. 400 CE, so later than Pappus) cannot be the First Principle (see Hadot [1978] 77-118, and [1990b] and [1993]), but in the present context this issue is not crucial. For the distinction between 'Demiurge' and 'Maker' in Plotinus, and the various hypostatic levels to which these terms are appplied in the Enneads see Charrue (1978) 123-39 (esp. on the interpretative echoes of Ti. 28c), O'Brien (1992) 331 n. 76.

³⁶⁴ Origen's title is quoted Porph. VP 3. For Origen the Platonist see Weber (1962), who collects the fragments and argues that he is not to be identified with the Christian. For 'Father' in Middle Platonism see Whittaker (1981). For 'King' as designation of the highest principle (lacking in Alcinous) see Dörrie (1970), whose interpretion of Origen's title is criticized by O'Brien (1993), who however fails to take Ti. 28c (for which see above, n. 357) into account. Alexander of Aphrodisias in Met. 59.29-31 Hayduck links Ti. 28c with Ep. 2.312e, and states that the first passage pertains to the efficient and the second to the final cause (Alexander's words are quoted Ascl. in Met. 52.21 ff. Hayduck). One should not forget that Plotinus was accused of plagiarizing Numenius (e.g. Porph. VP 17), and that (supposing the interpretation I attempt to argue is correct) his triad of primary hypostases cannot have been acceptable to Origen. O'Brien (1993) collects evidence to prove that Origen, in his turn, was criticized by Plotinus, which is plausible enough. I note in passing that Philoponus, having converted to Christianity, interpreted Ti. 28c—the formula 'Maker and Father' had been snapped up by many Christian authors before him—in the most simple way (Aet. 139.20 ff. Rabe), and interprets the King of [Plato] Ep. 2 as pertaining to the God who creates the cosmos (Aet. 645.1 ff.)

A view quite similar to Origen's is found in the Platonist Alexander of Lycopolis, a minor philosopher who is not very popular with the students of Neoplatonism. This man worked and had his own school at Alexandria around 300 CE, that is to say about a generation before Pappus.[365] Like Origen the Platonist (Fr. 7 Weber *ap.* Procl *Theol. Plat.* 2.4, esp. 2.31.8-11 Saffrey and Westerink) he moreover held the supreme principle to be an Intellect (*Contr. Man.* 10.4 Brinkmann, πρὸς ἐκεῖνον τὸν Νοῦν).

Pappus' evidence concerning *the* philosophers is quite at home in this company, and confirms that in the first part of the 4th century CE one could refer to a current view which, according to the assumptions of some contemporary scholars, had gone out of fashion centuries ago. The view cited by Pappus was one of the available options, and Christian authors—such as for instance the great and influential Athanasius of Alexandria (295-373 CE, so presumably a generation to a generation and a half younger than Pappus)[366]—who seem to have appealed to this variety of Platonism, were by no means as conservative, or as out of touch, as they sometimes have been believed to be. What these people did was, simply, to choose from among the available alternatives a view that was compatible with their particular purpose. And in the present

---

[365] Alex. *Contra Man.* 3.5-7 Brinkmann, τὸ ποιητικὸν αἴτιον ['efficient cause', a much less ambiguous term than ποιητής] τιμιώτατον τίθενται [*scil.*, the Christians] καὶ πρεσβύτατον καὶ πάντων αἴτιον τῶν ὄντων, a view which εἰκότως ἅπαντες ἂν ἀποδέξαιντο, and the detailed exposition at 9.21-10.4. The Christian God is among other things the 'Demiurge' of the universe (see e.g. also above, n. 362 on Philoponus), so Alexander's ποιητικὸν αἴτιον applies to this demiurgic function as well. For ποιητικὸν αἴτιον in this sense cf. e.g. Alex. Aphr. *in Met.* 34.6-8 Hayduck, μαρτυρεῖ δὲ Ἐμπεδοκλεῖ ὡς πρώτῳ τε διελόντι ποιητικὸν αἴτιον καὶ ταῖς ὑλικαῖς ἀρχαῖς καὶ στοιχείοις τοῖς τέτταρσι σώμασι κεχρημένῳ, 59.27-31, ζητήσαι δ' ἄν τις πῶς λέγοντος Πλάτωνος καὶ ποιητικὸν αἴτιον, ἐν οἷς λέγει "τὸν μὲν οὖν ποιητὴν καὶ πατέρα τοῦ παντὸς εὑρεῖν τε καὶ δεῖξαι ἔργον" κλτ. (on the King in *Ep.* 2.312e as final cause), and Simplius on Parmenides, *in Phys.* 34.14-6 Diels, Παρμενίδης ... ποιητικὸν αἴτιον ... μὲν ἓν κοινὸν τὴν ἐν μέσῳ πάντων ἰδρυμένην καὶ πάσης γενέσεως αἰτίαν δαίμονα τίθησιν. On this aspect of Alexander of Lycopolis' philosophy see Van der Horst and Mansfeld (1974) 10-3, on Alexander and Christianity Van der Horst (1996).

[366] E.g. Athan. *Contra gentes* 39.38-42 Thomson, 'because the creation is one, it is firmly believed that its Maker is also one. It is not the case that there is one cosmos because there is (only) one Demiurge, since God could also create other cosmoi. But since (only) one cosmos has come into existence, we have to believe that its Demiurge too is one (only)', ἑνὸς ὄντος τοῦ ποιήματος, εἷς καὶ ὁ τούτου ποιητὴς πιστεύηται. καὶ οὐχ ὅτι εἷς ἐστιν ὁ δημιουργός, διὰ τοῦτο καὶ εἷς ἐστιν ὁ κόσμος· ἠδύνατο γὰρ καὶ ἄλλους κόσμους ποιῆσαι ὁ Θεός. ἀλλ' ὅτι εἷς ἐστιν ὁ κόσμος ὁ γενόμενος, ἀνάγκη καὶ τὸν τούτου δημιουργὸν ἕνα πιστεύειν εἶναι. See Meijering (1996-8) 1.147, and cf. below, p. 130, complementary note 319.

case they did not even have to fall back upon a view that had long been dead.

One may of course wonder who *the* philosophers referred to by Pappus are. In the first place, I dare say, the most divine Plato himself. In the second place, without doubt, authors of introductions to Plato, and of Commentaries on Plato and Aristotle Pappus will have studied or even listened to, but whose works are lost. I feel in a position to submit this partial hypothesis because it has been shown on other occasions too that puzzling, or isolated, passages in an earlier author may be elucidated by what one finds in later authors.[367] As to the later commentators we shall find interesting explanations in Proclus, Philoponus and Simplicius. But we also have earlier evidence.

We must begin with the *fons et origo* of the discussion, a difficult passage in Plato's *Timaeus*. Plato says that the Demiurge gave the cosmos the shape of a sphere (my italics):[368]

> And for shape he gave it that which is fitting and akin to its nature,
> For the living being [i.e. the cosmos] that was to embrace all living

---

[367] See e.g. O'Meara (1989) 53-85 on the excerpts from the lost books of Iamblichus' *On the Pythagoreans*, and Mansfeld (1992) 243-62.

[368] *Ti.* 33b, σχῆμα δὲ ἔδωκεν αὐτῷ τὸ πρέπον καὶ τὸ συγγενές. τῷ δὲ τὰ πάντα ἐν αὑτῷ ζῷα περιέχειν μέλλοντι ζῴῳ πρέπον ἂν εἴη σχῆμα τὸ περιειληφὸς ἐν αὑτῷ πάντα ὁπόσα σχήματα· διὸ καὶ σφαιροειδές, ἐκ μέσου πάντῃ πρὸς τὰς τελευτὰς ἴσον ἀπέχον, κυκλοτερὲς αὐτὸ ἐτορνεύσατο, πάντων τελεώτατον ὁμοιότατόν τε αὐτὸ ἑαυτῷ σχημάτων, νομίσας μυρίῳ κάλλιον ὅμοιον ἀνομοίου; transl. Cornford, slightly modified. See Cornford (1937) 54, Vlastos (1975) 29, 94 n. 43. Cicero's translation, *Tim.* 17, "contains considerable additions"; see Pease (1955-8) 2.650; his translation of the formula I have italicized in the text is 'ea forma ... qua una omnes formae reliquae concluduntur'. At *Nat. deor.* 2.47, where the Platonic doctrine of the sphere is interpolated in the Stoic cosmology, he writes 'ea figura quae sola omnis alias figuras complexa continet' (Pease *ad loc.* cites a few parallels, but not the Pappus passages discussed here) Apul. *de Plat.* 1.198 paraphrases 'operiens omnia coercensque contineat'; Beaujeu (1973) 262 comments: the "monde, qui contient la totalité des réalités sensibles", so has missed the mathematical point. In the spurious Timaeus Locrus 208.5-8 Marg Plato's passage becomes εὖ δ' ἔχει καὶ καττὸ σχῆμα καὶ καττὰν κίνασιν, καθ' ὃ μὲν σφαῖρα ὄν, ὡς ὅμοιον αὐτὸ αὑτῷ παντᾷ εἶμεν καὶ πάντα τἆλλα ὁμογενέα σχάματα χωρῆν δύνασθαι, καθ' ἂν δὲ ἐγκύκλιον μεταβολὰν ἀποδιδὸν δι' αἰῶνος. Needless to say neither Cicero nor Apuleius provides a mathematical proof of this affirmation. The addition of 'good motion' in Timaeus Locrus (εὖ δ' ἔχει ... καττὰν κίνασιν) should be compared with the term εὐκινητότατον in the text of Alcinous quoted below, n. 370. Baltes (1972) 20-6 convincingly argues that Timaeus Locrus should be understood in the context of Middle Platonism and that the tract is a sort of mix, viz. part excerpt of the *Timaeus*, part interpretative additions from a *Timaeus* commentary. Perhaps even from more than one?

beings within itself the fitting shape would be *the figure that comprehends in itself all the figures there are*; accordingly, he turned its shape rounded and spherical, equidistant everywhere from centre to extremity—a figure the most perfect and uniform of all; for he judged uniformity to be ten thousand times more beautiful than its opposite.

It is the formula 'the figure that comprehends in itself all the figures there are',[369] stated without proof, which suggests to the mathematician the given that the isoperimetric regular solids can be inscribed in a sphere, and that accordingly this sphere comprehends them all and is the greatest of them all, that is to say has the greatest volume. Plato's undiluted eulogy of the sphere contains a correct mathematical definition (it is 'equidistant everywhere from centre to extremity'), so it is only natural to assume that also the formula 'the figure that comprehends in itself all the figures there are' has a mathematical connotation. But note that Plato means 'all living beings', viz. animals, men, and gods.[370] Animals and men evidently do not exhibit regular shapes in the mathematical sense of the word (it does not help to argue that they are compounds of such shapes, as Xenocrates seems to have done in a verbatim fragment attributed to him by Simplicius[371]). As to the gods, I would not know for certain what shapes to attribute to them: spheres, perhaps?

---

[369] One wonders whether Plato wanted to emend a doctrine attested (in Diogenes Laërtius, to be sure; derivation from Theophrastus, though defended by Diels, uncertain) for Leucippus, D. L. 9.31-2 = Leuc. 67A1 DK. Here we read that a cosmos comes into existence whenever in a big empty space numerous bodies (atoms) of all sorts of shapes (σώματα παντοῖα τοῖς σχήμασιν) come together. A spheroid compound is then formed, which forms a kind of membrane comprehending in itself all sorts of bodies (καὶ ποιεῖν πρῶτόν τι σύστημα σφαιροειδές. τοῦτο δ' οἷον ὑμένα ἀφίστασθαι περιέχοντα ἐν ἑαυτῷ παντοῖα σώματα). This account is quite different in this respect from the Atomist doctrine at ps.Plu. *Plac.* 1.4 (~Aët. 1.4 Diels), attributed to Leucippus also (67A24 DK) but probably later.

[370] This is analogous to the contents of the paradigm, *Ti.* 31a, 'that which embraces all the intelligible living creatures that there are', τὸ ... περιέχον πάντα ὁπόσα νοητὰ ζῷα. Same analogy at *Ti.* 30c-d: 'it (viz. the Living Being) embraces and contains within itself all the intelligible living beings, just as this universe embraces ourselves and all the other living beings that are visible', τὰ γὰρ δὴ νοητὰ ζῷα πάντα ἐκεῖνο ἐν ἑαυτῷ περιλαβὸν ἔχει, καθάπερ ὅδε ὁ κόσμος ἡμᾶς ὅσα τε ἄλλα θρέμματα συνέστηκεν ὁρατά (*scil.,* ἐν ἑαυτῷ περιλαβὸν ἔχει). Finally, on the cosmos, *Ti.* 69c: 'this universe, a single living creature containing in itself all the living creatures, mortal and immortal', πᾶν τόδε ..., ζῷον ἓν ζῷα ἔχον τὰ πάντα ἐν ἑαυτῷ θνητὰ ἀθάνατά τε.

[371] Quoted below, p. 131, complementary note 357.

Also note the other properties Plato ascribes to the sphere, viz. perfection and uniformity; these are instances of what Pappus calls 'natural characteristics'.

The so-called *Handbook* of Alcinous contains a cosmology which is an updated abstract from the *Timaeus*. The passage quoted above is here summarized as follows:[372]

> By way of shape, he bestowed on it sphericity, seeing as that is the fairest of shapes and the most voluminous and the most mobile.

The term πολυχωρότατος, 'having the greatest volume', is a mathematical *terminus technicus*. Clearly, Plato's a shade opaque formula 'that comprehends in itself all the figures there are' is interpreted in an acceptable mathematical way, though Alcinous too declines to provide a proof. The supreme mobility of the sphere is an Aristotelian ingredient, brought in in the wake of creative interpretation. Whittaker *ad loc.*[373] refers to *Ti.* 56a3 and 7, the only places in Plato where the *word* occurs; but Plato uses it not of the sphere but of the tetraeder, i.e. the extremely mobile element fire. Aristotle, on the other hand, who never uses this word, argues at *De cael.* 2.4.287a23-6 that the uniform movement of the outermost sphere is the fastest movement there is, that the fastest uniform movement is the shortest there is, so has to be circular. Therefore the heaven must be spherical. This fastest movement is not found in Plato, at least not explicitly; he argues that the heavens move in a circle because this is the *best* of all possible movements (*Ti.* 34e).

In the Commentary of Alexander of Aphrodisias on Aristotle's *Topics* we have a dialectical argument which beyond doubt contains a reference to the formula in the *Timaeus* (my italics):[374]

---

[372] *Did.*12.3, 167.46-168.2 Hermann, σχῆμα δ' αὐτῷ περιέθηκε τὸ σφαιροειδές, εὐμορφότατον σχημάτων καὶ πολυχωρότατον καὶ εὐκινητότατον. Transl. Dillon, slightly modified.

[373] In his *apparatus superior*. Note that this extra ingredient is lacking in the paraphrase of the *Timaeus* passage at Apul. *de Plat.* 198; for the parallel in Timaeus Locrus see above, n. 368. We should also refer to Arist. *An.* 1.2. 405a10-3, where we read that according to Democritus the soul consistst of very small fire atoms, which are the most mobile because they have the form of a sphere; also see Them. *in An.* 9.9-19 Heinze, Philop. *in An.* 67.12 f. Wallies. Both Democritus on the spherical atom and Plato on the fire element as being the most mobile are sharply criticized at Arist. *Cael.* 3.8.306b32-4 and 307a3-8, cf. Simpl. *in Cael.* 662.9 ff. Heiberg. But at [Arist.] *Mech.* 951b16-7 round shapes are said to be more mobile than others; also cf. e.g. Them. *in Phys.* 208.26 Schenkl, τὸ γὰρ σφαιροειδὲς εὐκίνητον γέγονεν.

[374] *in Top.* 76.9-15 Wallies, οἷον ὅτι ἀΐδιος ὁ κόσμος ἢ ὅτι σφαιροειδής. ἐπιχειρῆσαι γὰρ ἄν τις διαλεκτικῶς εἰς τοῦτο ὅτι τῷ τελειοτάτῳ τῶν σωμάτων οἰκεῖον τὸ

[...] for example that the cosmos is eternal, or spherical. One may try out the following dialectical argument about this: the most perfect shape is suitable for the most perfect of bodies; the cosmos is the most perfect of bodies, *for it contains all others in itself;* so the most perfect shape is suitable for the cosmos; now the sphere is the most perfect of shapes, for it admits neither addition nor subtraction;[375] accordingly the spherical shape is suitable for the cosmos.

In the updated excerpt from the *Timaeus* found in Diogenes Laërtius book III, in many ways different from its counterpart in Alcinous, we find another exegesis though one not entirely different from Alcinous':[376]

And it [*scil.*, the cosmos] is spherical because such is the shape of its Producer. For the latter contains the other Living Beings, and the former the shapes of them all.

Here the argument is from the product to the producer and back (a deduction on the basis of *Ti.* 29e3). The Demiurge, Diogenes says, contains the other Living Beings; this can only mean that the demiurgic Intellect[377] comprehends the prototypic Forms. And the spherical cosmos contains the shapes of all the living beings.

τελειότατον σχῆμα, ὁ δὲ κόσμος τελειότατον τῶν σωμάτων· πάντα γὰρ τὰ ἄλλα ἐν ἑαυτῷ ἔχει· τῷ κόσμῳ ἄρα τὸ τελειότατον τῶν σχημάτων οἰκεῖον· ἀλλὰ μὴν τελειότατον ἡ σφαῖρα τῶν σχημάτων· οὔτε γὰρ προσθήκην οὔτε ἀφαίρεσιν δέχεται· οἰκεῖον ἄρα τὸ σφαιρικὸν σχῆμα τῷ κόσμῳ.

[375] The formula οὔτε προσθήκην οὔτε ἀφαίρεσιν δέχεται occurs only here in Alexander. It is also found in Asclep. *in Met.* 310.20-4 Hayduck (with reference to ὥσπερ αὐτός φησιν ἐν τῇ Περὶ οὐρανοῦ) and 316.2-4, and Olymp. *in Mete.* 263.4-8 Stüve (also with reference to the *de Caelo*: ὡς ἐν τῇ Περὶ οὐρανοῦ πραγματείᾳ ἀποδείκνυσι). The *De caelo* passage is 2.4.286b18-25, though Aristotle here only says that the circle is perfect because it differs from the straight line in that there can be no πρόσθεσις to it, and that the same holds for the sphere; not a word about ἀφαίρεσις. Behind the fuller formula of Alexander and the Neoplatonists, we may believe, are two famous lines in Parmenides' description of the sphere, 28B844-5 DK, τὸ γὰρ οὔτε τι μεῖζον / οὔτε τι βαιότερον πελέναι χρεόν ἐστι τῇι ἢ τῇι, singled out for quotation by Plato *Sph.* 244c (who quotes three lines, 43-5), also quoted Procl. *in Parm.* 665.28-9 Cousin (who here starts at line 44 and omits τῇι ἢ τῇι) and in full, from Plato's *Sophist*, at *Theol. plat.* 3.20, 3.70.6-9 Saffrey and Westerink; the text is quoted from the *Sophist* too at Simpl. *in Phys.* 52.24-8 Diels (γέγραπται δὲ ἐν Σοφιστῇ τάδε κτλ.), quotation of the three lines being repeated *ibid.* 89.22-4 (the whole of B8 DK up to line 52, as is well known, is quoted *ibid.* 144.29 ff.) Two lines, B8.43-4, are quoted more or less paraphrastically at [Arist.] *MXG* 976a8-11, and the three lines again, B843-5 (without Parmenides' name) at Stob. *Ecl. phys.* 1.14.2. They were quite famous.

[376] D. L. 3.72, σφαιροειδῆ δὲ διὰ τὸ καὶ τὸν γεννήσαντα τοιοῦτον ἔχειν σχῆμα. ἐκεῖνον μὲν γὰρ περιέχειν τὰ ἄλλα ζῷα, τοῦτον δὲ τὰ σχήματα πάντων.

[377] The Demiurge is called an Intellect D. L. 3.69.

This passage looks like a predecessor of the more sophisticated fourth argument of Iamblichus (out of ten) in favour of the sphericity of the cosmos in his lost Commentary on the *Timaeus*, an abstract of which has been preserved by Proclus. I quote the passage:[378]

> Again, in addition to this, as the Intellible Living Being comprehends all the Intelligible Living Beings in one Unity, so the cosmos, in its assimilation to the Prototype, contains all the encosmic shapes by reason of its spherical shape; for only the sphere can include all the elements. Therefore, as by its singleness it reflects its similarity to the Intelligible All, so by its sphericity it imitates that All's containing of the wholes.

'All the elements': that is to say four of Plato's five regular convex solids. Iamblichus appears to be unaware of Archimedes' discovery that there are more such figures, or simply chooses to ignore it. This in spite of the fact that, as is clear from the passage in Heron quoted above, earlier exegetes of Plato (or so I presume) had argued that Plato already 'knew' two of Archimedes' solids and so, in principle, all of them. On the other hand, the 'Platonic figures', as they came to be called (perhaps to distinguish them from those of Archimedes),[379] are regular, whereas those of Archimedes are semi-regular. For this distinguishing characteristic of the 'Platonic figures' see Heron's description, *Def.* 103: 'these five are the *only* ones to be comprehended by equals [in size] and sames [in shape]'[379a]; *later* they were were given the name "Plato's

---

[378] Procl. *in Tim.* 2.72.31-73.3 Diehl = Iambl. Fr. 49 Dillon (whose transl. I have slightly modified): ἔτι πρὸς τούτοις ὡς τὸ νοητὸν ζῷον πάντα περιέχει τὰ νοητὰ ζῷα κατὰ μίαν ἕνωσιν, οὕτω καὶ ὁ κόσμος πρὸς τὸ παράδειγμα ὡμοιωμένος πάντα περιέχει τὰ ἐγκόσμια σχήματα κατὰ τὸ σφαιρικὸν σχῆμα· σφαῖρα γὰρ μόνη δύναται πάντα τὰ στοιχεῖα περιλαμβάνειν.

[379] First in Heron's reference to earlier authors, quoted below in the text (also cf. Heron *Metr.* 2.15, τῶν πέντε σχημάτων τῶν Πλάτωνος καλουμένων). The formula is rare; cf. further Procl. *in Eucl.* 68.22-3 Friedlein, τῶν καλουμένων Πλατωνικῶν σχημάτων, *Schol. Eucl.* XI 15, τὰ Πλάτωνος σχήματα, *Schol. Eucl.* XIII.1, ἐν τούτῳ τῷ βιβλίῳ, τουτέστι τῷ ιγ', γράφεται τὰ λεγόμενα Πλάτωνος ⟨ε⟩ [addition perhaps unnecessary] σχήματα, ἃ αὐτοῦ μὲν οὐκ ἔστιν [...], τὴν δὲ προσωνυμίαν ἔλαβεν Πλάτωνος διὰ τὸ μεμνῆσθαι αὐτὸν ἐν τῷ Τιμαίῳ περὶ αὐτῶν κτλ.

[379a] For this combination cf. e.g. Eucl. *Elem.* XI, hor. dem. 10, ἴσα δὲ καὶ ὅμοια στερεὰ σχήματά ἐστι τὰ ὑπὸ ὁμοίων ἐπιπέδων περιεχόμενα ἴσων τ'' πλήθει καὶ τ'' μεγέθει, XII dem. 3.87-8, ὑπὸ γὰρ ἴσων καὶ ὁμοίων ἐπιπέδων περιέχονται, XII dem. 8.23, τὰ ΒΗΜΛ, ΕΘΠΟ ἄρα στερεὰ ὑπὸ ὁμοίων ἐπιπέδων ἴσων τὸ πλῆθος περιέχεται; Heron *Def.* 116, διαφέρει μὲν καὶ ἐν στερεοῖς καὶ ἐν ἐπιπέδοις, ἤδη δὲ καὶ ἐν γραμμαῖς, ὁμοιότης καὶ ἰσότης; [Plu.] *Plac.* 879F , the heavenly bodies ὅμοια μὲν ἀνατέλλει τοῖς χρώμασιν, ἴσα δὲ τοῖς μεγέθεσι; *Schol. Eucl.* XI.5, οἷον εἰ στερεὸν σχῆμα περιέχεται φέρε εἰπεῖν ὑπὸ ⟨δ⟩ τριγώνων καὶ ⟨θ⟩ τετραγώνων καὶ τριῶν πενταγώνων, ἔτι δὲ καὶ

figures" by the Greeks [i.e., this became their standard designation in Greek]', εἰσὶ πέντε ταῦτα μόνον ὑπὸ ἴσων καὶ ὁμοίων περιεχόμενα, ἃ δὴ ὑπὸ τῶν Ἑλλήνων ὕστερον ἐπωνομάσθη Πλάτωνος σχήματα (for the distinction also compare Philoponus, e.g. *Aet.* 531.26 ff. Rabe). Anyhow Iamblichus sticks to the four solids constitutive of fire, earth, water and air. His exegesis of Plato's somewhat opaque formula, though more specific and outspoken, is to some degree still on the level of that of Alcinous, Alexander of Aphrodisias, and (the source of) Diogenes Laërtius. On the other hand it is also evident that it has been incorporated into a full-fledged Neoplatonic system. For this reason I believe that Pappus, when speaking of *the* philosophers, did not have his older contemporary Iamblichus in mind. For one thing, it is entirely uncertain whether he knew Iamblichus' work, while for another we may exclude Iamblichus because his view of the principles and their functions is far more complicated than that described by Pappus. Unless, of course, one recklessly assumes that Iamblichus said something simpler in a work for freshmen we no longer have.

We may finally cast a brief glance at the late commentators, and begin with Proclus. In his Commentary on the *Timaeus* he devotes quite a long section to the explanation of Plato's formula.[380] He argues that Plato's statement can be proved in three ways, viz. a philosophical way, a physicalist way, and a mathematical way. The physicalist arguments derive from Aristotle; I shall not discuss them. The mathematical argument is in two parts, an astronomical part which I leave to one side, and a mathematical part which I shall not translate but paraphrase.[381] Proclus correctly

---

ἕτερον στερεὸν σχῆμα ὁμοίως περιέχεται ὑπὸ ⟨δ⟩ τριγώνων καὶ ⟨θ⟩ τετραγώνων καὶ ⟨γ⟩ πενταγώνων ὁμοίων πάντων τοῖς προειρημένοις, ὅμοιά ἐστι τὰ στερεά, εἰ δὲ μὴ μόνον ὑπὸ ὁμοίων ἴσων τὸ πλῆθος περιέχεται ἑκάτερον, ἀλλὰ καὶ ἴσων, ἴσα τε καὶ ὅμοια κληθήσεται.

[380] Procl. *in Tim.* 2.68.7-76.29 Diehl. Compare Damascius' appeal to *Ti.* 31b at *in Phaed.* vers. 1.516, p. 261 Westerink (and vers. 2, p. 351), which according to Westerink *ad loc.* is "a selection from the comprehensive account given by Pr. *Tim.* II 68.14-76,29". This is correct, though Damascius varies the formulas, saying of the sphere that it is πανδεχέστατον (1.516.8) and μάλιστα πάντων χωρητική (2.117.3).

[381] Procl. *in Tim.* 2.76.7-29 Diehl, ὅτι δὲ καὶ ἡ σφαῖρα πολυχωρότατον τῶν ἰσοπεριμέτρων, ἀποδείκνυται παρ᾽ αὐτοῖς, καὶ ὅπως πάντα μὲν εἰς τὴν σφαῖραν ἐγγράφειν δυνατόν, οὐ πάντα δὲ εἴς τι τῶν πολυέδρων. καὶ οὐδὲν δεῖ μεταγράφειν ἡμᾶς τὰ παρ᾽ ἐκείνοις ἀποδεδειγμένα· πρὸς γὰρ τὸν δι᾽ ἐκείνων ἱκανῶς πεπαιδευμένον ποιούμεθα τοὺς λόγους· τοσοῦτον δὲ ὅμως ἱστορητέον, ὅτι τῶν ἰσοπλεύρων τε καὶ ἰσογωνίων καὶ ἴσην περίμετρον ἐχόντων τὸ πολυγωνότερον μεῖζον ἀποδείξαντες πρῶτον καὶ τὸν κύκλον ἐξῆς μείζονα τῶν ἰσοπλεύρων καὶ ἰσογωνίων, ἰσοπεριμέτρων δέ,

provides the gist of the mathematical proofs, and he admits that according to the experts there are more isoperimetric regular convex solids than are described by Plato in the dialogue. He continues by referring his readers, whom he takes to be sufficiently versed in mathematics, to the works of Euclid and Archimedes for the details. The latter, as will be remembered, is also mentioned by Pappus, who discusses his findings at the required length. Archimedes is simply absorbed by Proclus, and any criticism that could follow from a comparison between his stance and Plato's is in this way neutralized in advance.[382] Proclus then promises that for those especially interested he will add an appendix, or corollarium, 'after the whole treatise'. This is to contain the *sunagôgè*—in this sense a *hapax* in Proclus; think of Pappus' title—of all the mathematics one needs to understand the dialogue. This *sunagôgè* we do not have, and one may doubt that it was ever written. Apparently, inserting the complicated and lengthy mathematical proofs in the body of the Commentary would have been *ultra morem*.[383]

Philoponus deals with this issue in his Commentaries on Aristotle's *Analytica posteriora, De anima,* and *Physica,* Simplicius in his Commentaries on the *Physica* and *De caelo.* The commentators are prompted to do so by two kinds of passages in Aristotle. In several of his treatises Aristotle argues that there is a difference in competence between the practitioner of a particular science and the philosopher. At *Phys.* 2.4.193b23 ff., for instance, he speaks of the difference between physics and mathematics (read: mathematical astronomy), and submits that the physicist and the astronomer may deal with the same topics, e.g. whether the cosmos has the shape of a sphere, but will do so from a different point of view. In

δεικνύουσι καὶ τὴν σφαῖραν τῶν ἴσην ἐπιφάνειαν ἐχόντων στερεῶν σχημάτων ἑπομένως μείζονα καὶ διαφερόντως τῶν παρὰ Πλάτωνι λεγομένων πολυέδρων ἰσοπλεύρων καὶ ἰσογωνίων, τὰ μὲν χρώμενοι τοῖς παρὰ τῷ Εὐκλείδῃ δειχθεῖσι, τὰ δὲ τοῖς παρὰ τῷ Ἀρχιμήδει. καί, ὅπερ ἔφην, ἔξεστιν ἐκείνοις συγγενόμενον τὰς ἀποδείξεις ἀναλέγεσθαι· τάξομεν δὲ αὐτὰς καὶ ἡμεῖς ἐν τῷ μετὰ πᾶσαν τὴν πραγματείαν ἔχοντι τὴν συναγωγὴν τῶν πρὸς τὸν Τίμαιον μαθηματικῶν θεωρημάτων διὰ πλατυτέρων ἐφόδων ὧν τοῖς ὑπομνήμασιν ἐγκατασπείροντες γράφομεν, ἵν' ἐξῇ τοῖς φιλοθεάμοσι καὶ τούτοις ἔχειν ἠθροισμένα πάντα πρὸς τὴν τοῦ διαλόγου τῶν μαθηματικῶν ἕνεκα παντοίαν κατάληψιν. τῶν μὲν οὖν μαθηματικῶν ἅλις.

[382] In the later in *Euclid.* I, where Proclus argues that the aim of Euclid's *Elements* is the description and proof of the construction of the five Platonic solids (the 'cosmic figures' and their inscription in the sphere, 70.18 ff. Friedlein), not a word is said about Archimedes' discovery.

[383] For a similar attitude in Eutocius cf. above, n. 157.

his comments on this passage Simplicius refers to what he calls
the 'physical' arguments of Aristotle in the *De caelo* and those of
Plato in the *Timaeus*; by calling also Plato's argument 'physical' he
neutralizes in advance the kind of criticism mathematicians
would formulate, but this is by the way. Nevertheless, he adds that
the mathematical astronomer uses the given that the sphere is the
greatest of all isoperimetric figures.[384] More information, as
already intimated, is to be found in the *De caelo* itself, where the
sphericity of the heavens, the heavenly bodies, and the earth is
argued in a number of ways. One of these arguments has already
been cited: the uniform motion of the outer heaven is the fastest,
therefore the shortest, therefore circular, therefore the heaven is a
sphere. In his Commentary Simplicius this time refers to the
mathematical proofs of the proposition that the sphere is the figure
with the greatest volume.[385] On the one hand, he submits, these
proofs were already known before Aristotle's time, because he
presupposes them, while on the other these matters have been
expounded at length by Archimedes and Zenodorus. Here we
have Archimedes again, as in Proclus. A little later he says that
this thesis is Platonic and was accepted by Aristotle. For Plato had
said that the sphere is able to comprehend all the shapes—an
unmistakeable reference to the formula in the *Timaeus*. Virtually
the same arguments are to be found in Philoponus, including a
reference to Plato's formula.[386]

From these expositions in the Neoplatonic commentators, which
I have abridged rather drastically, it will be clear that these
passages in Aristotle and especially Plato, which so to speak cry out
to be explained, were at the focus of a discussion in which a variety
of traditional arguments were opposed to, or linked up with, each

---

[384] *in Phys.* 290.19-21 and 291.13-20 Diels .
[385] *in Cael.* 412.6-17 and 414.12-7 Heiberg, esp. διότι δέδεικται καὶ πρὸ
'Αριστοτέλους μὲν πάντως, εἴπερ αὐτὸς ὡς δεδειγμένῳ συγκέχρηται, καὶ παρὰ
'Αρχιμήδους καὶ παρὰ Ζηνοδώρου πλατύτερον, ὅτι τῶν ἰσοπεριμέτρων σχημάτων
πολυχωρητότερός ἐστιν ἐν μὲν τοῖς ἐπιπέδοις ὁ κύκλος, ἐν δὲ τοῖς στερεοῖς ἡ σφαῖρα,
and καὶ τοῦτο τὸ ἐπιχείρημα Πλατωνικὸν ὂν ὁ 'Αριστοτέλης ἠσπάσατο. εἰ γὰρ τῶν
ἐμβαδῶν ἴσων ὄντων ἐλαχίστη ἐστὶν ἡ τὸ κυκλικὸν περιέχουσα γραμμὴ καὶ διὰ τοῦτο
ἐλαχίστη, ὅτι τῶν ἰσοπεριμέτρων πολυχωρητότερος ὁ κύκλος, ὅπερ ὁ Πλάτων ἐδήλωσε
διὰ τοῦ περιέχειν πάντα ὁπόσα σχήματα.
[386] For instance *in An.* 56.4-21, esp. ὥσπερ ὁ Πλάτων ἐν τῷ Τιμαίῳ ἐζήτησε, διὰ τί
σφαιρικὸς ὁ οὐρανός· ὅτι, φησίν, ἔδει τὸ πάντων γενησόμενον δεκτικὸν καὶ περιέξον τὰ
πάντα τὸ πολυχωρητότατον τῶν σχημάτων σχήσειν· πολυχωρητότατον δὲ ἐν μὲν
ἐπιπέδοις ὁ κύκλος, ἐν δὲ στερεοῖς ἡ σφαῖρα, and *ibid.* 139.5-9 Hayduck.

other. Some of these references may seem to us somewhat far-
fetched in the particular contexts in which they occur. This attests
the importance they had acquired during centuries of exegesis and
discussion. Presumably Pappus too is acquainted with the ingredients of this discussion, which originated from Plato's formula in
the *Timaeus*. He generously accepts what Proclus was to call the
philosophical proofs, but submits that from his mathematical point
of view the argument of the philosophers is in part not a proof but a
mere allegation. It is true enough, but this truth is blind. He also
intimates that the divine Plato was insufficiently informed about
the number of regular solids that can be proved to exist, and these
criticisms purportedly also apply to other philosophers who are
insufficiently familiar with Euclid and especially Archimedes.
Proclus meets this critique (or rather a similar critique, for we may
very much doubt that he knew Pappus' *Sunagôgè*)[387] by strengthen-
ing the purely philosophical arguments. The heavy artillery of
Neoplatonic metaphysics is brought to bear on this matter, and he
creates a venerable tradition which leads up to Plato by quoting the
early philosophers as well as the ancient poets (and the poets he
believes to be ancient) on the sphere.

We next should look at the second of our passages in the
Sunagôgè. In book II chs. 12-7 Pappus deals with three kinds of
proportions between three quantitites, viz. the arithmetical, the
geometric, and the harmonic mean.[388] In ch. 18 he continues (my
italics):[389]

> Since Nicomachus the Pythagorean and some others have treated
> not only the first three proportions, which are most *useful*[390] for the
> *study* of the ancients, but also three others one finds with the
> ancients, and (since) in addition to these six (proportions) younger

---

[387] See above, n. 19 and text thereto.

[388] In all cases it holds that a > b > c. Arithmetical proportion: a - b = b - c,
e.g. 3 - 2 = 2 - 1. b, instantiated here as 2, is in the middle; this is why such
proportions are also callel *mesotètes*, means. Geometric proportion: a divided
by b = b divided by c, e.g. 4/2 = 2/1. Harmonic proportion: the quantity in the
middle is 1/3 of the first smaller than the first and 1/3 of the last bigger
than the last, e.g. 6 : 4 : 3. See further e.g. Etienne and Roels (1986).

[389] 1.84.1-8 Hultsch, ἐπεὶ δὲ καὶ Νικόμαχος ὁ Πυθαγορικὸς καὶ ἄλλοι τινὲς οὐ
μόνον περὶ τῶν πρώτων τριῶν μεσοτήτων [*scil.*, the arithmetical, geometric et
harmonic] εἰρήκασιν, αἳ χρήσιμοι τυγχάνουσιν μάλιστα πρὸς τὰς τῶν παλαιῶν
ἀναγνώσεις, ἀλλὰ καὶ περὶ ἄλλων τριῶν κατὰ τοὺς παλαιούς, καὶ ἐπὶ ταῖς ἓξ ταύταις
ἄλλαι ὑπὸ τῶν νεωτέρων προσεύρηνται τέσσαρες, πειρασόμεθα καὶ περὶ τούτων εἰπεῖν
ἐπιτονώτερον κτλ.

[390] See above, Ch. II 4.

authors have discovered four more (such proportions), we shall try
to speak of these too in a more thorough way ...

Pappus often refers to his predecessors, but he mentions names
only exceptionally.[391] This makes the present case all the more
worthwhile.[392] Heath says that Pappus "evidently despised" the
Introductio arithmetica,[393] but this remains to be seen. From the *Introductio* it is clear that Nicomachus is a Platonist rather than a
Pythagorean or, to formulate it differently: for Nicomachus Plato is
a follower of Pythagoras, so that he is able to place Plato's philosophy and even that of Plato's pupil Aristotle in a Pythagorean
succession—which evidently is a construct.[394] Strikingly enough,
this passage in Pappus is a mini-cento combining two passages of
Nicomachus himself. The 'study of the ancients' from the first
passage is combined with most of the contents of a passage in the
next chapter. The first of these passages runs:[395]

> After this it would be the proper time to incorporate the nature of
> proportions [i.e. combinations of ratios], a thing *most necessary*
> [issue of *utility*] for the study of nature and for the propositions of
> music, astronomy, and geometry, in particular for the *study in class
> of the ancients*.

---

[391] For Nicomachus' *Introductio arithmetica* see above, Ch. XI 1. Among
Pappus' 'others' we may perhaps include Theon of Smyrna, and the Pythagorean Moderatus of Gades (1st cent. CE); for the latter see Procl. *in Tim.* 1.19.4-6
Diehl, who when discussing the proportions mentions τοὺς Νικομάχους [...],
τοὺς Μοδεράτους καὶ εἴ τινες ἄλλοι τοιοῦτοι.

[392] For other references to secondary literature in Pappus see above, n. 39.
To the best of my knowledge this reference to Nicomachus has not been ex-
ploited by scholars, though Ver Eecke (1933) *ad loc.*, 1.63 n. 1 quotes *Ar.* 2.22.1.

[393] Heath (1921) 1.99.

[394] See above, Ch. XI 1 *ad fin.* It is no accident that Philolaus Fr. 44B12 DK
ap. Stob. *Ecl. phys.* 1.1.3, 18.5-7 Wachsmuth (for which see below, p. 130,
complementary note 319) was forged in order to prove that a prominent
Pythagorean had anticipated Plato. The same theory is attributed to
Pythagoras himself at Aët. 2.6.5 Diels, with the addition (ps.Plu. 887C only)
'Plato follows Pythagoras also as to this doctrine, Πλάτων δὲ καὶ ἐν τούτοις
πυθαγορίζει (this verb, first found in the comedians Cratinus, Antiphanes, and
Alexis is rare in prose; the present use is not paralleled earlier than Syr. *in
Met.* 22.21 Kroll, Πλατωνικοί γε ὄντες καὶ πυθαγορίζειν βουλόμενοι). These
attributions have the same background as the forgery attributed to Timaeus
Locrus, for which see above, n. 368.

[395] *Ar.* 2.21.1 at 119.19-22 Hoche, ἐπὶ δὲ τούτοις καιρὸς ἂν εἴη τόν περὶ ἀναλογιῶν
τ[ρ]όπον προσθέντας ἀναγκαιότατον ὄντα εἰς τὰς φυσιολογίας καὶ εἰς τὰ μουσικά τε καὶ
σφαιρικὰ καὶ γραμμικὰ θεωρήματα, οὐχ ἥκιστα δὲ καὶ εἰς τὰς τῶν παλαιῶν
συναναγνώσεις ... Transl. D'Ooge, modified. For these συναναγνώσεις see above.
n. 68, text to n. 184, n. 306 and text thereto. And cf. again above, Ch. XI 1 *ad
fin.*

The second, a little later, is as follows (my italics):[396]

> The *first three proportions*, then, which are acknowledged by all the ancients, Pythagoras, Plato, and Aristotle, are the arithmetic, geometric, and harmonic, *and there are three others* [...]; after which *the younger authors discover four more.*[397]

An account of the findings of these 'ancients' and 'younger authors' follows (*Ar.* 2.21.2-29 at 120.2-154.10, i.e. the last part of book II of the *Introductio*) which covers the same ground as the whole of Pappus *Coll.* 3.12-23 at 1.70.16-104.13 Hultsch, though in a different way. Nicomachus' exposition of the ten proportions is arithmetical, Pappus' proofs are geometric. It is, by the way, excluded that Pappus got his information about these sentences in Nicomachus via Iamblichus' revision of the *Introductio*, because the phrases I have italicized do not occur there.[398] Pappus did not despise Nicomachus; on the contrary, he found him most useful. His reference to Nicomachus *cum suis* moreover fits in nicely with his remark about *the* philosophers who hold that the First God is Demiurge. For this is also the point of view of Nicomachus in the *Introductio*; see 1.4.2 at 9.9-15 and 1.6.1 at 12.1-11 Hoche.[399]

From the passages in the *Sunagôgè* studied so far we may conclude that Pappus was in favour of and indebted to traditions of Platonic exegesis which, whether or not they called themselves Pythagorean, followed the letter of the *Timaeus* far more closely than some prominent Middle Platonists and the majority of the Neoplatonists did. It follows that, for a quite long time, varieties of

---

[396] *Ar.* 2.22.1 at 122.11-8 Hoche, εἰσὶν οὖν ἀναλογίαι αἱ μὲν πρῶται καὶ παρὰ τοῖς παλαιοῖς ὁμολογούμενα, Πυθαγόρᾳ τε καὶ Πλάτωνι καὶ Ἀριστοτέλει, τρεῖς πρώτισται ἀριθμητική, γεωμετρική, ἁρμονική, αἱ δὲ ταύταις ὑπεναντίαι ἄλλαι τρεῖς [...], μεθ᾽ ἃς καὶ ἄλλας τέσσαρας οἱ νεώτεροι εὑρίσκουσι ... Transl. D'Ooge, modified.

[397] Note that Proclus (in the passage cited above, n. 391) attributes the discovery of the proportions beyond the first three to Nicomachus *cum suis*, so cites Nicomachus cavalierly. A little later, *ibid.* 1.20.22-8 Proclus says that Nicomachus is right in calling the geometric proportion *analogia* and the others *mesotètes*. I note in passing that Philo, unlike Nicomachus, only knows and explains three proportions, viz. the first three, *Decal.* 20-1 (my italics): 'the decad contains *all* the analogies' (πάσας δ᾽ ἀναλογίας, τήν τε ἀριθμητικήν, ...καὶ τὴν γεωμετρικήν, ... ἔτι μέντοι καὶ τὴν ἁρμονικήν). Iambl. *in Nicom.* 100.15-24 Pistelli says that Pythagoras and his intimate followers (including Archytas) only knew three proportions (μόναι δὲ τὸ παλαιὸν τρεῖς ἦσαν μεσότητες ἐπὶ Πυθαγόρου καὶ τῶν κατ᾽ αὐτὸν μαθηματικῶν).

[398] So Iamblichus is not one of the 'others' (see above, n. 391).

[399] *Pace* D'Ooge *& al.* (1926) 108; better O'Meara (1989) 16.

Platonism must have coexisted which, on the basis of their metaphysics, one would be inclined to arrange diachronically.

Pappus returns to the mathematics of the *Timaeus* elsewhere in the *Sunagôgè*, viz. in same the chapter (3.18) where, as we have seen, he appeals to Nicomachus and others. This is the third passage I wish to discuss. Here we find a cento of reminiscences of ideas and formulas to be found all over the dialogue.[400] This, I believe, shows that Pappus here as well is indebted to an exegetical tradition: comments, or Commentaries, which adopted the method of explaining Plato from Plato, a variety of the better known *Homerum ex Homero* principle.[401] He argues as follows:[402]

> Since the geometric mean, which derives its primary origin from equality, establishes both itself and the other means, it indicates, as the most divine Plato says, that the nature of proportion is the cause of harmony for all things and of their reasonable and ordered coming into existence For he states that the divine nature of proportion is the single bond of all the mathematical disciplines,[403] and the cause of the coming into existence and the bond of all things that come to be. We shall demonstrate the constitution of the ten means through the geometric mean ...[404]

---

[400] Esp. *Ti.* 24c, 29e, 30a, 31b–32c (the main passage), 41e, 42c, 42e, 80b. The same phenomenon is found, on a more extensive scale to be true, in several passages in Alcinous, *Did.* ch. 12, who however on the whole tends to follow the drift of the exposition in the *Timaeus* more faithfully, as is only to be expected in an excerpt. But this passage is less technical where the mathematic side of things is concerned. On the geometric proportion in the *Timaeus* and what follows from it see e.g. Cornford (1937) 45-52.

[401] See Mansfeld (1994) 241, index *s.v.* interpretation.

[402] 1.86.19-88.4 Hultsch, ἡ τοίνυν γεωμετρικὴ μεσότης ἐκ τῆς ἰσότητος τὴν πρώτην λαβοῦσα γένεσιν αὐτή τε αὑτὴν καὶ τὰς ἄλλας συστήσει μεσότητας, ἐνδεικνυμένη, καθά φησιν ὁ θειότατος Πλάτων, τὴν τῆς ἀναλογίας φύσιν αἰτίαν τῆς ἁρμονίας πᾶσι καὶ τῆς εὐλόγου καὶ τεταγμένης γενέσεως· λέγει γὰρ ἕνα δεσμὸν εἶναι τῶν μαθημάτων ἁπάντων, αἰτία δὲ γενέσεως καὶ δεσμὸς πᾶσι τοῖς γενομένοις ἡ τῆς ἀναλογίας θεία φύσις. δειχθήσεται δὲ ἡ σύστασις τῶν δέκα μεσοτήτων διὰ τῆς γεωμετρικῆς ἀναλογίας ...

[403] For this meaning of μαθημάτων see above, n. 81, n. 196 *ad fin.*, Pappus himself, *Coll.* 2.636.11 οἱ ἀπὸ τῶν μαθημάτων ('mathematicians'), Plu. *Non posse* 1086CD, and cf. LSJ *s.v.* μάθημα 3: e.g. the three disciplines arithmetic, geometry, and astronomy (already in Plato) to which canonics was added later.

[404] Theon of Smyrna, *Util.* 106.12-9 Hiller, quoting Adrastus, also says that the other proportions (of which apparently a larger number, viz. 12, is assumed to exist) are dependent on the geometric mean: ἐπανιτέον δὲ ἐπὶ τὸν τῶν ἀναλογιῶν καὶ μεσοτήτων λόγον. μεσότητές εἰσι πλείονες, γεωμετρικὴ ἀριθμητικὴ ἁρμονικὴ ὑπεναντία πέμπτη ἕκτη. λέγονται δὲ καὶ ἄλλαι πάλιν ἓξ ταύταις ὑπεναντίαι. τούτων δέ φησιν ὁ Ἄδραστος μίαν τὴν γεωμετρικὴν κυρίως λέγεσθαι καὶ ἀναλογίαν καὶ πρώτην· ταύτης μὲν γὰρ αἱ ἄλλαι προσδέονται, αὐτὴ δ' ἐκείνων οὐχί, ὡς ὑποδείκνυσιν ἐν τοῖς ἐφεξῆς.

Pappus' references to the *Timaeus* are far richer than those of
Nicomachus in the latter's chapter about the geometric proportion,
though Nicomachus explicitly appeals to the passage in the
dialogue which in this respect is the most important[405] (Pappus'
reference to this passage, as we noted, is not stated *disertis verbis*).
What Pappus appears to have done, in a way entirely comparable
to Nicomachus' treatment of the same subject, is to interpret Plato's
statement as a programmatic injunction. An interpretation of this
nature obviously was already traditional. Pappus, therefore, provides proofs for all the means, or proportions, concerned, even
those discovered by 'younger authors' and so not to be found in the
'ancients'. What is more, he endorses the cosmological impact and
function of the proportional equality which is at the basis of the
mean and included in it, as argued by Plato, and accordingly
accepts the rational ordering of the cosmos which according to
Plato is produced thereby.

---

[405]   *Ar.* 2.24,6 at 129.14-9 Hoche, referring to *Ti.* 31c.

COMPLEMENTARY NOTES

COMPLEMENTARY NOTE 5 (to p. 2)

For references to literature on the late prolegomena to rhetorical treatises see Mansfeld (1994) 52-3. We now moreover have the correct and full text of Aelius Theon's *Progumnasmata*, including the final chapters lost in Greek and preserved in Armenian, splendidly edited by Patillon and Bolognesi (1997) with the assistance of other scholars such as J.-P. Mahé, L. Pernot and A. Ouzounian. I note here that in this early treatise too, to be dated to the 1st-2nd cent. CE, isagogical issues (or notions) are used as a matter of course to impart structure to the exposition (see overview at Patillon and Bolognesi [1997] xxiv-xxviii). Description of *subject* at the beginning, 59.13-6 Spengel, ἃ δὲ πρὸ τῆς ὑποθέσεως ἀναγκαῖόν ἐστιν εἰδέναι τε καὶ ἐπιεικῶς ἐγγυμνάζεσθαι, ταῦτα νῦν πειράσομαι παραδοῦναι, which at the same time shows that this introductory work, or *part* of rhetoric, comes before (πρό) another part, or treatise. Also cf. 61.26-9 where moreover the type of διδασκαλία—see below, p. 128, complementary note 217—is mentioned, and see further above, Ch. V 8. *Utility* : 60.1, 60.20, 60.27, 60.32, 61.5 Spengel etc., see Patillon and Bolognesi (1997) 223, index *s. vv.* χρήσιμος, ὠφέλεια, ὠφελεῖν, ὠφέλιμος. *Systematic arrangement*: e.g. 64.28 ff., τὴν δὲ τάξιν τῶν γυμνασμάτων αὐτῶν οὕτω ποιησόμεθα· πρῶτον μὲν ..., ἔπειτα δὲ κτλ.; see further *ibid.* 223, index *s. v.* τάξις. *Qualities to be expected of the teacher*: 65.29 ff. Spengel, πρῶτον μὲν ἁπάντων χρὴ τὸν διδάσκαλον ἑκάστου γυμνάσματος εὖ ἔχοντα παραδείγματα ἐκ τῶν παλαιῶν συγγραμμάτων ἀναλεγόμενον προστάττειν τοῖς νέοις ἐκμανθάνειν κτλ.

COMPLEMENTARY NOTE 11 (to p. 4)

For πρόθεσις ('project') in Diodorus Siculus, closely linked with the contents of the work as a whole and to those of the individual books as well as to the division of the work into books, see *Hist.* 1.52 (τῆς ὅλης προθέσεως), 1.98.10 (κατὰ τὴν ἐν ἀρχῇ τῆς βίβλου πρόθεσιν), 3.74.6, 4.85.7, 13.114.3, 14.117.9, 15.95.4, 16.95.5, 17.118.3, 18.75.3, 19.110.8. For Ptolemy cf. above, n. 237. Numerous examples of ὑπόθεσις ('subject', 'theme') in Dionysius of Halicarnassus, e.g. *Ant.* 1.1.2.8-9 (ὑποθέσεις ... καλὰς καὶ μεγαλοπρεπεῖς καὶ πολλὴν ὠφέλειαν τοῖς ἀναγνωσομένοις φερούσας), *Lys.* 15.15, *Lys.* 20.14 ff. where the hypothesis (like part of that preceding a play) is a brief summary, *Is.* 4.12, *Thuc.* 6.19-21, which moreover is also about unity and division (κατὰ τὸ λαβεῖν ὑπόθεσιν μήτε μονόκωλον παντάπασι μήτ' εἰς πολλὰ μεμερισμένην καὶ ἀσυνάρτητα κεφάλαια), cf. *ibid.* 7.13 ff. The word περιοχή ('abstract') too may come close to this meaning, cf. above, p. 45, while Eutocius uses σκοπός in the sense of πρόθεσις. Also compare Latin *periocha*, used e.g. for the abstracts from Livy and for the contents of plays by Plautus and Terentius. Ptol. *Geogr.* 1.2, 1.5.17 Nobbe uses τέλος, 'purpose' (cf. also above, n. 257). The subject, or rough—or even at first sight incomprehensible (see Mansfeld [1995], where also more comprehensible examples are discussed)—indication of the contents of a work as well as the identification (or at least indication) of its author are of course introductory *topoi* right from the start of Greek literature. For proems in Plato and their precedents in the philosophical and rhetorical traditions see Algra (1996) 47-51, Runia (1997) 103-11; also see Birt (1882) 464-81, van Sickle (1980) 7-8. For Aristotle's so-called esoteric works cf. Cic. *Ep. Att.* 4.16.2, 'in singulis libris utor prohoemiis ut Aristoteles in iis quos ἐξωτερικούς vocat'.

COMPLEMENTARY NOTE 26 (to p. 9)

For Analysis-and-synthesis in this difficult but highly important passage of Pappus (as well as in Greek mathematics in general) see Hintikka and Remes (1974), and Jones (1986a) 1.66-71 (who has missed Hintikka and Remes). Schrenk (1994) 97-8 leaves the solution of the difficulties of Pappus' description to "students of the history of mathematics". Crombie (1994) 1.276-309, "Analysis and Synthesis", places ancient mathematical Analysis (*ibid.* 282 ff.) in a broader context. The best study of Pappus on Analysis and synthesis in the context of Greek mathematics known to me is Mäenpää (1993) 139-200; also see the summary of his argument at Mäenpää (1997) 201-7. The few examples of theoretical analyses to be found are in areas where Greek geometry verged on algebra, e.g. book II of the *Elements*; see Knorr (1986) ch. 8. Note that Pappus in the *Collectio* sometimes only presents an analysis and omits the synthesis; see Hintikka and Remes (1974) 29; this only holds for problematic analysis, where the synthesis would be a trivial conversion. For Marinus on Pappus on Analysis above, text to nn. 208 and 209, and n. 219. For Apollonius' view see above, text to n. 126; Pappus talks about Analysis here, not synthesis, so there is no conflict.

I have found two parallels for Pappus' formula ἀνάπαλιν λύσις, viz. Elias *in Isag.* 37.21-3 Busse, οὐδὲν γάρ ἐστιν ἀνάλυσις, εἰ μὴ ἀπόδειξις ἀντεστραμμένη, ὅθεν καὶ ἀνάλυσις ὡς ἀνάπαλιν λύσις οὖσα τοῦ προκειμένου, and *Schol. vet. in Theocr.* 17.27 Wendel, ἀνάλυσις τὸ σχῆμα κατὰ φιλοσόφους· ἀνάλυσις δέ ἐστιν ἀντεστραμμένη ἀπόδειξις τουτέστιν ἀνάπαλιν λύσις. Note that these instances do not derive from a mathematical context.

COMPLEMENTARY NOTE 56 (to p. 19)

Gal. *Synopsis libr. De pulsibus* 9.455 Kühn; he also wrote a *De pulsibus ad tirones* (Περὶ σφυγμῶν τοῖς εἰσαγομένοις, 8.453 ff., which begins with the words "Οσα τοῖς εἰσαγομένοις ... χρήσιμον ἐπίστασθαι περὶ σφυγμῶν, ἐνταῦθα λεχθήσεται. τὴν δ' ὅλην ὑπὲρ αὐτῶν τέχνην ἑτέρωθι γεγραμμένην ἔχεις). The relation between the three treatises is expressed as follows 9.463: λεχθήσεται δὲ καὶ νῦν τὰ κατ' αὐτὸ χάριν τοῦ μηδὲν ἐλλείπειν τῶν ἀναγκαίων τῇ νῦν ἐνεστώσῃ πραγματείᾳ, ἀλλ' ἔχειν τοὺς φιλοπονεῖν βουλομένους ἐν ἐλαχίστῳ μὲν τὰ πρῶτα καὶ ἀναγκαιότατα κατὰ τὴν εἰσαγωγήν (scil., the Ad tirones), ἐν διεξόδῳ δὲ τελεωτάτῃ τὰ κατὰ τὴν μεγάλην πραγματείαν (scil., the De puls.), ἐν τῷ μέσῳ δ' ἀμφοῖν τὰ νῦν λεγόμενα (scil., the Synopsis). On the relation between the Eisagogè and the great treatise see also De libr. propr. ch. 5, 19 K. = Scr. min. 2.110.4-25 Mueller, where the treatise in seventeen books is again called the μεγάλη πραγματεία, and the other work is referred to as Περὶ χρείας σφυγμῶν τοῖς εἰσαγομένοις. For the titles of the introductory works cf. also above, text to n. 303. In his introductory treatise De musculis ad tirones (Περὶ μυῶν τοῖς εἰσαγομένοις) 18B.927 Galen refers to the De usu partium as follows: περὶ δὲ τῆς χρείας (scil., τῶν μυῶν) ἅμα τοῖς ἄλλοις ἅπασιν ἐν τῇ μεγάλῃ πραγματείᾳ τῇ περὶ χρείας μορίων (scil., εἴρηταί μοι). For the formula μεγάλη πραγματεία itself see also De const. artis med. 1.295, Anat. admin. 2.217, De meth. med. 11.145, and PHP 8.1.15 = Posid. Fr. 38 Edelstein-Kidd; this, pace Kidd (1988) 1.182, in view of Galen's usage must have been a multi-book treatise.

COMPLEMENTARY NOTE 67 (to p. 20)

Note that in Thrasyllus' tetralogical catalogue of Democritus' works at D. L. 9.46 the title Μέγας διάκοσμος (the first of the physics section) comes *before* Μικρὸς διάκοσμος, so the latter can hardly have been viewed as an introduction to the former. In this section of the catalogue the treatises are listed in a way which, though involving the order of study, enumerates them in a sequence of diminishing generality, not of increasing difficulty. The Μέγας διάκοσμος according to 'some' moreover is to be attributed to Leucippus (D. L., loc. cit.), so the adjectives μέγας and μικρός here not only serve to distinguish two different treatises dealing with the same subject but also two different authors, and the order apparently is according to the dates of these authors. As Pierluigi Donini suggests we may also compare the title α' ἔλαττον given to book II of Aristotle's Metaphysics to distinguish it from book A (with capital A). Here no order of study can be intended in the sense that the 'small' book comes before the 'big' one; the issue is that both books are a sort of introduction to the rest of the composite treatise, and that in later antiquity there was a discussion as to which of these alternatives is genuine, or that perhaps both are (see references in the apparatus of Jaeger's OCT ed. of the Metaphysics p. 33, though his conclusion is not good; see Berti [1982] and Vuillemin-Diem [1983]). Similar terminology is used to distinguish works with the same title ascribed to a single author, as in the case of the Ἀλκιβιάδης (and Ἱππίας) μείζων and ἐλάττων, Olymp. in Alc. § 3.6-7 (here no order of study, or systematic order, is involved); cf. above, n. 54 and text thereto, for *minor* and *major*. In Thrasyllus' catalogue of Plato at D. L. 3.50, however, the distinction is effected by numbering these dialogues. For the numbering of the titles Aristotle's two *Analytics* on the basis of their *themes, order of study*, and s *ystematic order* see Alex. *in APr.* 7.9-11 Wallies, ἐπεὶ τοίνυν πρότερον μὲν συλλογισμός, ὕστερον δὲ ἀπόδειξις, εἰκότως, ἐν οἷς μὲν βιβλίοις περὶ τοῦ προτέρου τὸν λόγον ποιεῖται, ταῦτα Πρότερα ἐπέγραψεν, ἐν οἷς δὲ περὶ τοῦ ὑστέρου, ταῦτα Ὕστερα (cf. *ibid.* 7.33-8.2, Ammon. *in APr.* 5.8-7.23 Wallies); also see Aristotle's catalogue at D. L. 5.23, Προτέρων ἀναλυτικῶν in eight books, Ἀναλυτικῶν ὑστέρων μεγάλων in two books, and 5.29 τὰ Ἀναλυτικὰ πρότερα καὶ ὕστερα. Another parallel (though not involving an order of study) is Elias' distinction, *in Isag.* 32.34-33.2 Busse, between [Aristotle's] *Magna moralia* and Aristotle's *Ethica Nicomachea* as, respectively, Μεγάλα Νικομάχεια and Μικρὰ Νικομάχεια; the odd reason given is that the former were addressed by Aristotle to his father Nicomachus and the latter to his son Nicomachus. Cf. *ibid.* 116.16-9, καὶ Νικομάχεια τά τε μικρὰ καὶ τὰ μεγάλα· τὰ μὲν γὰρ τῷ πατρὶ προσφωνεῖ Νικομάχῳ καὶ λέγονται Νικομάχεια μεγάλα, τὰ δὲ τῷ υἱῷ ὁμωνύμῳ τῷ πατρὶ καὶ λέγονται Νικομάχεια μικρά. The latter case however is to be explained by the length of the scrolls, see Birt (1882) 493-4. Birt gives further examples of this type of title: Ἰλιας μικρά (title e.g. Arist. *Poet.* 1459b2, Paus. 3.26.9, Clem. *Strom.* 1.21.104.2, but the 'little *Iliad*' in four books is small compared to the *Iliad*), Μεγάλα ἔργα, see Hes. Frs. 286-7 Merkelbach & West *ed. minor* ('big' presumably in comparison with what has been preserved as the Ἔργα καὶ ἡμέραι), Μεγάλαι ἠοῖαι see Frs. 246-62 Merkelbach & West *ed. minor* (title e.g. Athen. 8.66.16, Paus. 2.2.3. 2.16.4, 4.1.8; this epic presumably longer than the much similar Γυναικῶν κατάλογος *sive* Ἠοῖαι), and the Hippocratic τὸ δεύτερον Περὶ νούσων τὸ μεῖζον and τὸ μικρότερον, different in length. For Ptolemy's smaller *Fourbooks* as contrasted with his *Great Suntaxis* see above, Appendix 1.

COMPLEMENTARY NOTE 77 (to p. 23)

Tannery (1882) argued that Proclus knew Heron via Pappus, but the fact that his Commentary was still accessible to Anaritius (see above, n. 90 and text thereto) shows that it can hardly have been inaccessible to Proclus. References to Heron are at *in Euclid.* 196.15 ff. Friedlein (critical), 305.21 ff. (reference), 323.7 ff., (οἱ περὶ Ἥρωνα καὶ Πορφύριον, approvingly), 346.13 ff. (quotation), and 429.13 ff. (critical), but Van Pesch (1900) 121-2 on the basis of the material in Anaritius has proved that Proclus also uses Heron without mentioning his name. The same undoubtedly holds for his use of Pappus and others, but in some cases this may have been caused by interpolations from their Commentaries in the text (on interpolations from Heron's Commentary in the text of Euclid see Heiberg [1883-8] 2.564-7; for Pappus see above, n. 78; for Theon above, n. 87). It is generally assumed (e.g. also by Sezgin [1974] 153, "mit Sicherheit identisch") that the titles given under 'Heron' at *Fihrist* 7.2, Dodge (1970) 2.642: 'Book on solving the uncertainties of Euclid', and under 'Account of his [i.e. Euclid's] book on the Elements of Geometry' at *Fihrist* 7.2, Dodge (1970) 2.635: 'Heron explained this book, solving its uncertainties' (cf. Suter [1892] 22 and 16) refer to the Commentary, but I believe that it is not to be excluded that the book on 'solving the uncertainties' (or 'difficulties') is to be distinguished from the 'Account' (= Commentary) and either is to be identified with the original Τὰ πρὸ τῆς γεωμετρικῆς στοιχειώσεως (cf. above, Ch. VI 8), or pertains to a lost treatise belonging to the ἀπορήματα (or ζητήματα, or προβλήματα) καὶ λύσεις literature, for which see Gudeman (1927). Heiberg (1903) 58-9 shows that some of the *Scholia in Eucl.* too are derived from Heron.

COMPLEMENTARY NOTE 89 (to p. 26)

The fragments in Anaritius contain no matter of an introductory kind. Simplicius' Commentary is mentioned in the *Fihrist* ch. 7.2 under 'Simplicius al-Rumi', Dodge (1970) 2.60: 'Exposition of the beginning of the book of Euclid, which is an introduction to the art of geometry' (cf. Suter [1892] 21). See also Heath (1926) 1.27-8. Note that Simplicius refers to and quotes other books of the *Elements* as well in his long abstract from book II of Eudemus' *History of Geometry* (Fr. 140 Wehrli, pp. 59 ff.); ὀλίγα τινὰ προστιθεὶς ⟨εἰς⟩ σαφήνειαν ἀπὸ τῆς τῶν Εὐκλείδου Στοιχείων, as he says *in Phys.* 60.28-9 Diels. These additions are picked out from the whole work: from *Elem.* book I *in Phys.* 61.1 ff., 63.8 ff., 65.19 ff., from book II 62.9 ff., from book III 61.28 ff. (twice), 65.29 ff., 66.13 ff., 69.8 ff., from book IV 68.13 ff., and from book XII 61.9 ff. An abstract from book VI is found *in Phys.* 492.6 ff., a reference to Alexander of Aphrodisias on Euclid at 511.21 ff. Diels. He also refers to Euclid not by name but as ὁ στοιχειωτής (*in Cael.* 414.2 Heiberg), and uses Euclidean material without any reference at all as well. Overview of references to and quotations from Euclid in the commentators on Aristotle at Heiberg (1903) 352-4. I may perhaps add that Galen mentions Euclid's name eight times (including the reference in the *Timaeus* abstract).

COMPLEMENTARY NOTE 108 (to p. 30)

This also is an issue in another scholion which however does not belong to the oldest collection, viz. *Schol.* V.3: 'this book is said to be by Eudoxus of Cnidus, the mathematician who flourished in the times of Plato; yet it is ascribed to Euclid though not according to a false title (ἐπιγέγραπται δὲ ὅμως Εὐκλείδου, ἀλλ᾽ οὐ κατά τινα ψευδῆ ἐπιγραφήν). For insofar as the discovery is concerned there is nothing which hinders it from belonging to someone else, but in view of the sequential and systematic arrangement of the theorems (κατὰ στοιχεῖον ... συντάξεως) and of the fact of their being entailed by other theorems which are arranged in this way, it is agreed by all that is is by Euclid'. Cf. above, text to n. 117. The remark about Eudoxus recalls *Schol. vat.* V.1, see above, text to n. 107, and presumably indicates that this piece of information too goes back to an early Commentary. On the *Elements* in relation to earlier mathematical literature see further *Schol.* X.62, XII.12, XII.38, XIII.1, Procl. *in Eucl.* 68.6-11, and e.g. the overview in Lloyd (1973) 34-9.

COMPLEMENTARY NOTE 119 (to p. 34)

The Greek text of this passage in the Arabic Pappus runs as follows, *Schol.* X 1.70-9: ὅτι δὲ χρήσιμος ἡ τούτων θεωρία, μὴ καὶ περιττὸν λέγειν. τῶν γὰρ Πυθαγορείων λόγος τὸν πρῶτον τὴν περὶ τούτων θεωρίαν εἰς τοὐμφανὲς ἐξαγαγόντα ναυαγίῳ περιπεσεῖν, καὶ ἴσως ἠνίττοντο, ὅτι πᾶν τὸ ἄλογον ἐν τῷ παντὶ καὶ ἄλογον καὶ ἀνείδεον κρύπτεσθαι φιλεῖ [see below], καὶ εἴ τις ἂν ψυχὴ ἐπιδράμοι τῷ τοιούτῳ εἴδει τῆς ζωῆς πρόχειρον καὶ φανερὸν τοῦτο ποιήσηται, εἰς τὸν τῆς γενέσεως ὑποφέρεται πόντον καὶ τοῖς ἀστάτοις ταύτης κλύζεται ῥεύμασιν. τοιοῦτον σέβας καὶ οὗτοι εἶχον οἱ ἄνδρες περὶ τὴν τῶν ἀλόγων θεωρίαν. For the metaphors in τὸν τῆς γενέσεως ὑποφέρεται πόντον καὶ τοῖς ἀστάτοις ταύτης κλύζεται ῥεύμασιν compare Procl. *in Tim.* 1.113.29-31 Diehl, ὁ γὰρ Ἠριδανὸς ποταμὸς καὶ ἡ ἐκεῖ πτῶσις τὴν εἰς τὸν πόντον τῆς γενέσεως ἐνδείκνυται τῆς ψυχῆς φοράν, Olymp. *in Grg.* ch. 47.6 Westerink, ἰστέον ὅτι οἱ φιλόσοφοι τὸν βίον τὸν ἀνθρώπειον θαλάττῃ ἀπεικάζουσιν, Simpl. *in Phys.* 360.31-2 Diels, τὰ ἐν τῷ πόντῳ τῆς γενέσεως ... ὡς τὸ ἄστατον τῆς γενέσεως κατευθυνούσης, Iambl. *Myst.* 7.2, ἰλὺν μὲν τοίνυν νόει τὸ σωματοειδὲς πᾶν καὶ ὑλικὸν ἢ τὸ θρεπτικὸν καὶ γόνιμον ἢ ὅσον ἐστὶν ἔνυλον εἶδος τῆς φύσεως μετὰ τῶν ἀστάτων τῆς ὕλης ῥευμάτων συμφερόμενον, ἢ ὅσον τὸν ποταμὸν τῆς γενέσεως χωρεῖ, [Basil.] *Consol. aegr.* Migne *PG* 31.1717.7-9, τοιοῦτος ὁ τῶν ἀνθρώπων βίος, ἄστατος θάλασσα, ἀὴρ ἀνώμαλος, ὄναρ ἀβέβαιον, ῥεῦμα παρατρέχον, καπνὸς διαχεόμενος, σκιὰ μεταπηδῶσα, πέλαγος ὑπὸ κυμάτων ἐνοχλούμενον. For similar metaphorical language see already e.g. the influential passage Plato *Tht.* 152e (quoted Eus. *PE* 14.4.1 and Stob. *Ecl. phys.* 1.19.9); then Plot. *Enn.* 3.6.6, Simpl. *in Cat.* 354.27 Kalbfleisch, *in Phys.* 77.32-5, 789.19-20, and 1313.8-9 Diels. Heracliteanizing Platonism without Forms (see the passages collected at Marcovich [1978] 137-40; also above, n. 229), projected upon the Pythagoreans. This makes Pappus' use (not cited in Marcovich's edition of the fragments) of a Heraclitean formula, viz. Fr. B123 DK = 8 Marcovich, φύσις κρύπτεσθαι φιλεῖ, all the more interesting.

COMPLEMENTARY NOTE 192 (to p. 59)

See e.g. Neugebauer (1975) 2.893, Simon (1988) passim. Simon's contention (summarized at [1997] 193-6) that ancient optics is geometric and *psychological* rather than mathematical and physicalist is a trifle confusing, since according to the ancients psychology is a part of physics. Chrysippus' (and Apollodorus') doctrine of vision (which became the standard Stoic view, and one very much indebted to mathematical and mainstream Greek optics) is part of their psychology and therefore treated in the physical section in Diogenes Laërtius, viz. at 7.157 = *SVF* 2.867 and *SVF* 3 Apoll.12. That some Stoics made the theory of presentation and sense-perception a part of 'logic' to be treated before phonetics and semantics (see D. L. 7.41 and 48 ff., on which Mansfeld [1986] 356 and 361 ff.) is another matter. On the Stoic theory of vision see further the material collected by Ingenkamp (1971). The theory of vision of the Atomists is exceptional in that it has no room for the visual ray, but it, too, fails to grant light its proper role, cf. e.g. Simon (1988) 37-8, Crombie (1994) 1.156.

COMPLEMENTARY NOTE 217 (to p. 64)

This has to do with *brevity* or *fullness* and so comes close to the issue of *clarity*. Already mentioned (together with *utility*) in Galen's evaluation of earlier Commentaries on the Hippocratic *Aphorisms, in Aph.* 17B.351-2 Kühn, ὅσοι τοίνυν ἢ τοῦ *τρόπου τῆς διδασκαλίας* ἢ ὅλως *τῆς χρείας* τῶν συγγραμμάτων *αἰτίαν ἀποδίδοσθαι κατὰ τὸ προοίμιόν* φασιν [*scil.*, of the *Aphorisms*, the first of which, 'life is short' etc., is considered to be the proem], οὗτοί μοι δοκοῦσιν ἄμεινόν τι τῶν ἄλλων γινώσκειν. τό τε γὰρ ἀφοριστικὸν εἶδος τῆς διδασκαλίας, ὅπερ ἐστὶ τὸ διὰ βραχυτάτων ἅπαντα τὰ τοῦ πράγματος ἰδίᾳ περιορίζειν, χρησιμώτατον τῷ βουλομένῳ μακρὰν τέχνην διδάξαι ἐν χρόνῳ βραχεῖ. For the expression τρόπος τῆς διδασκαλίας and its implications see further Gal. *Anat. adm.* 2.236.2-3, 239.17, 240.19, *San. tuend.* 6.102.9-10, 347.15-48.1, *Dign. puls.* 8.947.17, *Meth. med.* 10.101.8-11, *in Aph.* 17B.355.9-10 Kühn, S.E. *P.* 3.266, Clem. Alex. *Strom.* 6.8.64, Epict. *Diss.* 2.14.2.4, Iambl. *VP* § 20, Procl. *in Parm.* 1027.27-9 Cousin. For the later commentators see e.g. Amm. *in Isag.* 23.17-9 Busse, Philop. *in Cat.* 27.25-7 Busse, *in APo.* 3.14 Wallies, *in An.* 227.25 Hayduck, Elias *in Isag.* 41.27-8 Busse, David *in Isag.* 80.13 and 95.9-10 Busse; see also above, n. 5, and Mansfeld (1994) 23.

COMPLEMENTARY NOTE 225 (to p. 67)

By the time of Pappus' Commentary chapter divisions and headings were in the text, but at least for book V these are sometimes different from those in the Ptolemy mss. (For the headings in Iamblichus as probably his own see O'Meara [1989] 35 with n. 14). The same holds for the Commentary of Theon of Alexandria (but see above, text to n. 266), composed about thirty years later (note that the still later commentator of whom fragments are extant in a Parisian manuscript [above, n. 222] speaks *disertis verbis* of the meaning of the ἐπιγραφὴ ... τοῦ β΄ κεφαλαίου τοῦ ζ΄

βιβλίου of the *Suntaxis*; text at Tihon [1976] 183). See Toomer (1984) 5, who cites the evidence for Pappus and Theon and argues that "Ptolemy himself did not use any chapter divisions at all"; so in his translation he brackets all chapter headings. The issue however is not as clear-cut as that, see e.g. Rome (1931) 48 n. 1; one should moreover also look at Ptolemy's other works (for the *Apotelesmatica* see above, Ch. XI 2, for the *Harmonica* Düring [1930] lxxvi, who argues that the headings are beyond doubt genuine), and at the practice of other authors. This does not entail that headings underwent no change in the course of transmission; for a possible case see above, n. 243. Useful notes with references to the literature at Saffrey and Westerink (1968) 1.129 n. 2, and Haase (1982) 121 n. 313. Also see Petitmingin (1997) which however is mostly on the evidence concerning tables of contents, not chapter headings in the works themselves, in Latin literature. The medieval mss. of the *Placita* of ps.Plutarch (ca. 150 CE, the author thus being a contemporary of Ptolemy's) do have chapter headings; in fact this treatise cannot dispense with them, and they are confirmed for its 1st cent. CE source Aëtius by Stobaeus. In the tiny papyrus fragments (early 3rd cent.) from Antinoöpolis of ps.Plutarch's *Placita* there is room for four chapter-headings, though actually in only one case a part of such a heading is extant; see Mansfeld and Runia (1997) 127. For the chapter headings of Quintilian see Mutschmann (1911) 96-7; for those of the shared source of Sextus *Pyrrh. Hyp.* and the first section of ps.Galen *Hist. philos.* see *ibid.* 97-8.

COMPLEMENTARY NOTE 260 (to p. 76)

Proclus wrote an entire treatise, the *Hypotyposis astronomicarum positionum* ed. Manitius (1909), extant, mostly dealing with the astronomy of Ptolemy (thought he also mentions other names) from a philosophical point of view, and he often refers to Ptolemy elsewhere. The *Hypotyposis* is not a Commentary on the *Megalè Suntaxis* however, so does not come within our present scope. On Proclus and astronomy see e.g. Neugebauer (1975) 3.1036, Segonds (1987), Siorvanes (1996) 262-311. To his pupil and successor Marinus Ptolemy was the best guide for this discipline (Dam. *Isid. ap.* Phot. *Bibl.* cod. 242.145 [p. 198 Zintzen], ὁ ἄριστος ἡγεμὼν Πτολεμαῖος τῆς ἀστροθεάμονος ἐπιστήμης), cf. also above, n. 222. Ammonius Hermiae taught Ptolemy's astronomy (to Damascius, see Dam. *Isid. ap.* Phot. *Bibl.* cod. 181, 126b Bekker [Zintzen p. 191], ... ἐξηγητὴν αὐτῷ γεγενῆσθαι Δαμάσκιος ἀναγράφει καὶ τῆς συντάξεως τῶν ἀστρονομικῶν Πτολεμαίου βιβλίων), which helps to explain the numerous references to Ptolemy in Philoponus' editions of Ammonius' commentaries, and in Simplicius. He also lectured on the astrolabe (also discussed by Proclus in the *Hypotyposis*), see above, n. 313. His astronomical tables are extant in Arabic, see Endress (1987) 405. On Pappus and Theon in this context see also Neugebauer (1975) 2.965-8.

COMPLEMENTARY NOTE 308 (to p. 87)

See further O'Meara (1989), who argues the importance of Nico-machus—via—Iamblichus for our understanding of certain strands of subsequent Platonism, though it does matter a little that he fails to refer to Hippolytus (see above, n. 307). On Iamblichus' free version of Nico-machus' treatise *ibid.* 51-2. The *Introductio* is less often referred to by the Neoplatonists (even Iamblichus in the *in Nicom.* rarely mentions his name) than one would perhaps expect. Syrian. *in Met.* 103.6-8 Kroll mentions Nicomachus together with Iamblichus, ἐντυχὼν οὗτος ταῖς τε Νικομάχου συναγωγαῖς τῶν Πυθαγορείων δογμάτων καὶ ταῖς τοῦ θείου Ἰαμβλίχου περὶ αὐτῶν τούτων πραγματείαις, and *ibid.* 151.18-21 refers to him and other Pythagoreans. Proclus cites him twice, *in Tim.* 1.19.4 and 20.26 Diehl, the first time in the company of other Pythagoreans (see above, n. 391). Simplicius refers to Nicomachus and Iamblichus together, *in Cael.* 507.14 Heiberg, Νικόμαχος καὶ Νικομάχῳ κατακολουθῶν Ἰάμβλιχος. David *Prol.* 26.9 ff. Busse has a verbatim quotation from the first chapter of the *Introductio* but finds it necessary to explain who Nicomachus is (εἷς δὲ οὗτος τῶν Πυθαγορείων); by calling him a Pythagorean he disagreed with Ammonius *cum suis* (see above, text to n. 314), perhaps however not on purpose. References which do not apply to the *Introductio*: three in Porphyry, viz. *VP* §§ 20 and 59 (in Iambl. *VP* there is only one named reference, viz. at § 251), and *Contra Christ.* fr. 39.32 Von Harnack *ap.* Eus. *HE* 6.19.8, where he is listed among the pagan philosophers said by Porphyry to have been studied and followed by Origen the Christian.

COMPLEMENTARY NOTE 319 (to p. 90)

Philo *Opif. mund.* 171 argues from the unicity of the cosmos to the unicity of God, see Runia (1986) 174-5. For Athanasius' argument see above, n. 366. For Alcin. *Did.* ch. 12, 167.41-3 Hermann the unicity of the cosmos derives from that of the Paradigm (ἰδέα); cf. Calcidius *in Tim.* 276.14-77.9 Waszink. An extensive argument is found at Procl. *in Tim.* 2.68.21 ff. Diehl, where Proclus gives what he calls Plato's threefold philosophical proof of the sphericity of the cosmos (see above, Appendix 2). The first of these is 'from the One' and involves the unicity of the Demiurge, of the Paradigm, of the Good, and of the sphere: αὐτίκα ἀπὸ τοῦ ἑνὸς εἴποις μὲν ἄν, ὅτι καὶ ὁ δημιουργὸς εἷς, εἴποις δ' ἄν, ὅτι καὶ τὸ παράδειγμα ἕν, εἴποις δ' ἄν, ὅτι καὶ τὸ ἀγαθὸν ἕν ἐστιν, καὶ ἀπὸ τούτων ἂν λάβοις, ὅτι καὶ ἐν τοῖς σχήμασι τὸ μάλιστα ἓν τοῦ μὴ ἑνὸς θειότερόν ἐστι καὶ τελειότερον. ὃ γάρ ἐστιν ἐν ⟨τοῖς⟩ θεοῖς τὸ ἕν, καὶ ὃ ἐν τοῖς νοητοῖς ζῴοις τὸ ἐν αὐτοζῴον, καὶ ὃ ἐν τοῖς δημιουργοῖς ὁ εἷς ποιητὴς καὶ πατήρ, τοῦτο ἐν τοῖς σχήμασι τοῖς στερεοῖς ἡ σφαῖρα.

COMPLEMENTARY NOTE 357 (to p. 105)

The term 'Maker' is also found in the paraphrastic introduction (also = Philol. Fr. 44A13 DK, though the reference to Philolaus is questionable) to a fragment of Speusippus (Fr. II 4 Lang = Fr. 28 Tarán) *ap.* [Iambl.] *Theol. Ar.* 83.5 De Falco: τῷ τοῦ παντὸς ποιητῇ θεῷ. Without going into the vexed question of the sources of the *Theologumena arithmetica* (tortuous discussion at Tarán [1981] 291-8, who tends to ascribe too much to Iamblichus), one may agree with Tarán's note on the formula at issue, *ibid.* 272, viz. that the 'image of god as the creator of the universe' was 'probably' taken by Speusippus 'from the *Timaeus*' (i.e. 28c). Also see Huffman (1993) 362: 'clearly the Platonic demiurge'. If it was not Speusippus himself who said this, it will have been his excerptor (perhaps Nicomachus, see Huffman [1993] 361).

In my view there is no sufficient reason to doubt the correctness of the summary of Speusippus, although the equation of the five regular solids with the five 'cosmic *elements*' is unplatonic. Plato's dodecaeder, though inscribable in a sphere and even being capable of being blown up to form a sphere (above n. 348), is not an element. The formula used in the Speusippus abstract seems to presuppose the Aristotelian aether (spherical) as fifth, or first, element. But Speusippus' interpretation is paralleled in a fragment of Xenocrates, where the aether is also said to be one of the five Platonic elements: Fr. 53 Heinze = Frs. 265-6 Isnardi Parente *ap.* Simpl. *in Phys.* 1165.3 ff. Diels and *in Cael.* 12.22 ff. Heiberg (I quote the second of these texts, italics mine): ... Ξενοκράτης ὁ γνησιώτατος αὐτοῦ τῶν ἀκροατῶν ἐν τῷ Περὶ τοῦ Πλάτωνος βίου τάδε γράφων· "τὰ μὲν οὖν ζῷα οὕτω διῃρεῖτο εἰς ἰδέας τε καὶ μέρη πάντα τρόπον διαιρῶν, ἕως εἰς τὰ *πέντε στοιχεῖα* ἀφίκετο τῶν ζῴων, ἃ δὴ πέντε *σχήματα καὶ σώματα* ὠνόμαζεν, εἰς *αἰθέρα* καὶ *πῦρ* καὶ *ὕδωρ* καὶ *γῆν* καὶ *ἀέρα*". This doctrine is also found in the certainly spurious fragment Philol. 44B12 DK, see Burkert (1972) 276, Huffman (1993) 392-5. Kraus and Walzer (1951) 60 (*in appar.*) are not entirely correct. I also believe, *pace* Huffman and others, that [Philolaus'] four 'bodies *in* the sphere' (τὰ ἐν τᾶι σφαίραι [*scil.*, σώματα]), viz. fire, water, earth and air, contrasted with the 'rotating sphere' (as I translate the formula ὃ τᾶς σφαίρας ὁλκάς [mss.], or ὁλκόν [Burkert], of this phoney Doric) as a fifth body (πέμπτον), must be four of Plato's five regular solids, which can be inscribed *in* it, the sphere itself of course being a regular solid too. For σώματα as regular solids see the excerpt from Iamblichus quoted above, n. 377, and for the equivalence of σχήματα and σώματα see the Xenocrates fragment quoted above. On the issues involved also see Moraux (1963) 1182-4, 1187, 1192-3.

BIBLIOGRAPHY

Standard editions both in the fields of ancient mathematics and astronomy and more recherché standard editons in the fields of ancient philosophy have been listed, because the former are probably less familiar to historians of philosophy, and the latter to historians of mathematics. Moreover, very few mathematical and astronomical texts are available on the D version of the TLG, and in some cases only Arabic translations are extant. The ancient authors are to be found below under the name of their editor(s); the names of the authors concerned are listed as well, followed by references to these editions and to the more obvious secondary literature. The numbers after commentaries, texts with commentaries or notes, monographs, or papers refer to the location of my citations in the footnotes (abbreviated n, e.g. 5n4 means p. 5 note 4), in the complementary notes (abbreviated cn, e.g. 180cn5 means p. 180 complementary note 5), and in the text (just the page number), and so may serve as an *Index nominum modernorum.*

Aelius Theon see Patillon and Bolognesi (1997)
Alcinous see Whittaker and Louis (1990), Dillon (1993)
Alexander of Aphrodisias see Hayduck (1891), Wallies (1883), Wallies (1891)
Alexander of Lycopolis see Brinkmann (1895), Van der Horst and Mansfeld (1974), Villey (1985), Van der Horst (1996)
Algra, K. A. (1996) 'Observations on Plato's Thrasymachus: the case for *pleonexia*', in Algra *& al.* (1996) 41-60   123cn11
Algra, K. A., Van der Horst, P. W. and Runia, D. T., eds. (1996) *Polyhistor: Studies in the History and Historiography of Ancient Philosophy*, Philos. Ant. 72 (Leiden/New York/Cologne)
Allatius, L. (1731) *Procli Paraphrasis in Ptolemaei Libros IV. De Siderum Affectionibus*, with Latin transl. (Leiden)   81n284
an-Nayrizi see Anaritius
Anaritius see Besthorn and Heiberg (1893-1932), Curtze (1899), Tummers (1984), Tummers (1994)
Angeli, A. and Colaizzo, M. (1979) 'I frammenti di Zenone Sidonio', *Cronache Ercolanesi* 9, 47-133   23n76
Angeli, A. and Dorandi, T. (1987) 'Il pensiero matematico di Demetrio Lacone', *Cronache Ercolanesi* 17, 89-103   23n76
Apollonius of Perga see Halleius (1706), Halleius (1710), Balsam (1861), Heiberg (1891-3), Heath (1896), Toomer (1970), Toomer (1990)
Apuleius see Beaujeu (1973), Moreschini (1991)
Archimedes see Heiberg (1910-5), Dijksterhuis (1956), Mugler (1972)
Arrighetti, G. (^{2}1973) *Epicuro. Opere*, with Italian transl. & notes, Bibl. cult. filos. 41 (Turin)   106n361
Asclepius of Tralles see Hayduck (1888), Tarán (1969)
Athanasius see Thomas (1971), Meijering (1996-8)
Attalus see Maass (1898)
Atticus see Des Places (1977)
Aujac, G., ed. (1975) *Géminos. Introduction aux phénomènes*, with introd., French transl. & notes, Coll. Budé (Paris)   23n79

Aujac, G., ed. (1979) *Autolycos de Pitane. La sphère en mouvement, Levers et couchers héliaques, Testimonia*, avec la collab. de Brunet, J.-P. et Nadal, R., with introd., French transl. & notes, Coll. Budé (Paris)   15n45

Autolycus of Pitane see Mogenet (1950), Aujac (1979)

Balsam, H., transl. (1861) *Des Apollonius von Perga sieben Bücher über Kegelschnitte nebst dem durch Halley wieder hergestellten achten Buche* (Berlin)

Baltes, M. (1972) *Timaios Lokros. Über die Natur des Kosmos und der Seele*, kommentiert von M. B., Philos. Ant. 21 (Leiden)   109n368

Bardy, G., ed. (1952-8) *Eusèbe de Césarée. Histoire ecclésiastique* 3 vols., with French transl. & notes, SC 31, 41, 55 (Paris)

Barnes, J. (1975) *Aristotle's Posterior Analytics*, transl. with notes   83n295

——, (1997) *Logic and the Imperial Stoa*, Philos. Ant. 75 (Leiden/New York/Cologne)   5n12a

Beaujeu, J., ed. (1973) *Apulée. Opuscules philosophiques (Du Dieu de Socrate, Platon et sa doctrine, Du monde) et fragments*, with French transl. & notes, Coll. Budé (Paris)   106n359 106n360

Bellosta, H. (1997) 'Ibrahim ibn Sinan: Apollonius arabicus' in Hasnawi & al. (1997) 32-48   10n29

Berthelot, M. and Ruelle, C. E., eds. (1888) *Collection des anciens alchimistes grecs* t. 1 *Introduction*; t. 2 *Texte grec*; t. 3 *Traduction* (Paris, repr. Osnabrück 1967)   94n330

Berti, E. (1982) 'Note sulla tradizione delle due primi libri della <<Metafisica>>', *Elenchos* 3, 5-38   124cn67

Bertier, J., transl. (1978) *Nicomaque de Gérase. Introduction arithmétique*, with introd., notes & ind., Hist. Doctr. Ant. Class. 2 (Paris)   n300

Besthorn, R. O. and Heiberg, J. L., eds. (1893-1932) *Euclidis Elementa, ex interpretatione al-Hadschdschadsii cum commentariis al-Nairizii*, arabice et latine 6 vols. (Copenhagen)

Birt, T. (1882) *Das antike Buchwesen in seinem Verhältnis zur Litteratur, mit Beiträgen zur Textgeschichte des Theokrit, Catull, Properz und anderer Autoren* (Berlin, repr. Aalen 1959)   69n238 123cn11

Boer, A. and Weinstock, S., eds. (1940) *Porphyrii philosophi Introductio in Tetrabiblum Ptolemaei*, in Weinstock, S., ed., *Catalogus Codicum Astrologorum Graecorum* vol. 5.4 (Brussels) 185-228

Bogaert, P.-M. (1997) 'Eptaticus : le nom des premiers livres de la Bible dans l'ancienne tradition chrétienne grecque et latine', in Fredouille & al. (1997) 313-37   97

Boll, F. (1894) 'Studien über Claudius Ptolemäus. Ein Beitrag zur Geschichte der griechischen Philosophie und Astrologie', *Jahrbb. Class. Phil.* Suppl. 21, 49-244   66n226 67n227 71n247 72n249

Boll, F. and Boer, A., eds. (1940) *Claudii Ptolemaei Opera quae exstant omnia* vol. 3.1, *Apotelesmatica*, Bibl. Teubn. (Leipzig, repr. with corr. 1957)

Brinkmann, A. (1895) *Alexandri Lycopolitani Contra Manichaei opiniones disputatio*, Bibl. Teubn. (Leipzig, repr. Stuttgart 1989)

Brummer, J., ed. (1912) *Vitae vergilianae* (Leipzig; repr. Stuttgart 1969)

Bulmer-Thomas, I. (1971) 'Eutocius of Ascalon', in Gillispie (1970-90) 4.488-91   40n133

——, (1974) 'Pappus of Alexandria', in Gillispie (1970-90) 10.293-304   6n15 32n114 94n331

Burkert, W. (1972) *Lore and Science in Ancient Pythagoreanism* (Cambridge MA) 25n86 29n104 29n107 32n114 32n115 33n118 131cn357

Busse, A., ed. (1900) *Eliae (olim Davidis) in Aristotelis categorias commentarium*, CAG 18.1, 107-255 (Berlin)

——, ed. (1902) *Olympiodori prolegomena et in Categorias commentarium*, CAG 12.1 (Berlin)

Busse, A., ed. (1904) *Davidis prolegomena et in Porphyrii isagogen commentarium*, CAG 18.2 (Berlin)

Calcidius see Waszink (1962)

Cantor, M. (^{3}1907) *Vorlesungen über die Geschichte der Mathematik* Bd. 1: *Von den ältesten Zeiten bis zum Jahre 1200 n. Chr.* (Leipzig, repr. New York 1965) 61n197

Cassiodorus see Mynors (1937)

Charrue, J.-M. (1978) *Plotin lecteur de Platon* (Paris)   107n363

Cherniss, H., ed. (1976) *Plutarch's Moralia* vol. 13.1, *999C–1032F*, Loeb Cl. Libr., text of *Plat. Quaest., De an. procr.* and of the *epitome* of the latter, with English transl. & notes (Cambridge MA/London)

Chiaradonna, R. (1997) rev. Mansfeld (1994), *Elenchos* 18, 158-65   2n4

Clagett, M. (1978) *Archimedes in the Middle Ages* vol. 3: *The Fate of the Medieval Archimedes* pt. 3: *The Medieval Archimedes in the Renaissance, 1450-1565* (Philadelphia)   31n113

Cleomedes see Todd (1990)

Cornford, F. M. (1937 and later repr.) *Plato's Cosmology. The Timaeus of Plato translated with a running commentary* (London)   105n357 109n368 120n400

Crombie, A. C. (1994) *Styles of Scientific Thinking in the European Tradition. The history of argument and explanation especially in the mathematical and biomedical sciences* 3 vols. (London)   24n79 50n168 59n191 59n194 100n338 123cn26 128cn192

Curtze, M., ed. (1899) *Anaritii in decem libros priores Elementorum Euclidis commentarii ex interpretatione Gherardi Cremonensis, in codice Cracoviensi 569 servata* = *Euclidis opera omnia*, Suppl. (Leipzig)

D'Ooge, M. L., transl. (1926) *Nicomachus of Gerasa. Introduction to Arithmetic.* With Studies in Greek Arithmetic by Robbins, F. E. and Karpinski, L. C. (New York, repr. New York/London 1972)   82n287 85n300 89n318 119n399

Damascius see Zintzen (1967), Westerink (1977)

David see Busse (1900), Busse (1904)

De Falco, V., ed. (1922) *[Iamblichi] Theologumena arithmeticae*, Bibl. Teubn. (Leipzig, repr. with add. by Klein, U., Stuttgart 1975)

De Lacy, Ph. H. (1978-84 and later repr.) *Galen. On the Doctrines of Hippocrates and Plato*, I-II: *Edition, Translation*, III: *Commentary and Indices, CMG* V 4,1,2 (Berlin)

De Libera, A. and Segonds, A.-Ph. (1998) *Porphyre. Isagoge*, Greek & Latin *Lesetext*, with French transl., introd. & notes, Coll. Sic et Non (Paris)   56n185 73n254

Decorps-Foulquier, M. (1992) 'L'époque où vécut le géomètre Sérénus d'Antinoé', in Guillaumin, J.-Y., ed., *Mathématiques dans l'Antiquité*, Mém. Centre Jean-Palerme 11 (Saint-Étienne) 51-8   3n8 93n325

——, (1997) 'L'édition d'Eutocius d'Ascalon des Coniques d'Apollonius de Perge: un exemple du rôle des écoles de l'antiquité tardive dans la transmission des textes scientifiques grecs', in Hasnawi *& al.* (1997) 49-60   41n133, 42n141

Des Places, É., ed. (1966) *Jamblique. Les mystères d'Égypte*, with French transl. & notes, Coll. Budé (Paris)

——, ed. (1973) *Numénius. Fragments*, with French transl., comm. & notes, Coll. Budé (Paris)   101n340 106n362

——, ed. (1977) *Atticus. Fragments*, with French transl., comm. & notes, Coll. Budé (Paris)   101n340

Detel, W., transl. (1993) *Aristoteles Analytica Posteriora*, with comm. = Aristoteles Werke in deutscher Übersetzung Bd. 3 T. II.1-2 (Berlin/Darmstadt)   83n295

Devreesse, R. (1954) *Introduction à l'étude des manuscripts grecs* (Paris)   1n1
    7n17 38n128
Deubner, L., ed. (1937) *Iamblichus. De vita pythagorica liber*, Bibl. Teubn.
    (Leipzig, repr. with add. & corr. by Klein, U., Stuttgart 1975)
Dicks, D. R. (1972) 'Geminus', in Gillispie (1970-90) 5.344-7   24n79
Diehl, E., ed. (1903-6) *Procli Diadochi in Platonis Timaeum commentaria* 3 vols.,
    Bibl. Teubn. (Leipzig, repr. Amsterdam 1965)
Diels, H., ed. (1882) *Simplicii in Aristotelis Physicorum libros quattuor priores
    commentaria* 2 vols., CAG 9.1-2 (Berlin)
——, (1893) 'Ueber das physikalische System des Straton', *SB Ak. Berlin*, 101-
    27, repr. in Burkert, W., ed. (1969) *Hermann Diels. Kleine Schriften zur
    Geschichte der antiken Philosophie* (Darmstadt) 239-65   49n166 50n167
    50n168
Diels, H. and Schramm, E., eds. (1918) 'Herons Belopoiika (Schrift vom
    Geschutzbau)', griech. & deutsch, *Abh. Ak. Berlin*, Phil.-hist. Kl. 1918 Nr.
    2 (Berlin, repr. Leipzig 1970)
Dijksterhuis, E, J. (1929-30) *De Elementen van Euclides*. 1: *De ontwikkeling der
    Grieksche wiskunde voor Euclides. Boek I der Elementen;* 2: *De boeken II-XIII der
    Elementen,* 2 vols. in 3 (Groningen)
——, (1956) *Archimedes*, Act. hist. sc. nat. & med. 13 (Copenhagen, repr. with
    bibliogr. suppl. 'Archimedes after Dijksterhuis' by Knorr, W. R.,
    Princeton 1987)   48n158
Dillon, J. M. (1973) *Iamblichi Chalcidensis in Platonis dialogos commentariorum
    fragmenta,* ed. with English transl. & comm., Philos. Ant. 23 (Leiden)
    113n378
——, (1977) *The Middle Platonists. A Study of Platonism 80 B.C. to A.D. 220*
    (London, rev. ed. with new afterword 1996)   82n287 99n337 100n339
——, transl. (1993) *Alcinous. The Handbook of Platonism*, with introd. & comm.
    (Oxford, repr. 1995)   99n337
Diocles see Toomer (1976a)
Dodge, B., transl. (1970) *The Fihrist of al-Nadim: A Tenth-Century Survey of
    Muslim Culture*. Records of Civilization: Sources and Studies 83 2 vols.
    (New York)   96, 126cn77
Donini, P. L. (1969) 'Il sublime contro la storia nell'ultimo capitolo del ΠΕΡΙ
    ΥΨΟΥΣ', *Parol. pass.* 126, 190-202   72n250
——, (1982) *Le scuole, l'anima, l'impero. La filosofia antica da Antioco a Plotino,*
    Sintesi 3 (Turin)   72n250 82n287 99n337 100n339 101n340
Dorandi, T., ed. (1991a) *Filodemo. Storia dei filosofo. [.] Platone e l'Academia,*
    Scuola di Epicuro 12 (Naples)   6n14
——, (1991b) 'Den Autoren über die Schulter geschaut. Arbeitsweise und Auto-
    graphie bei den antiken Schriftstellern', *Zeitschr. Pap. Epigr.* 87, 11-33
    38n128
——, (1994) 'La tradizione papiracea degli "Elementi" di Euclide', in *Proceed.
    20th Intern. Congr. Papyrol.* (Copenhagen) 307-11   25n87
——, (1997a) 'Tradierung der Texte im Altertum; Buchwesen', in Nesselrath,
    H. G., ed., *Einleitung in die griechische Philologie* (Stuttgart/ Leipzig) 1-16
    7n17 16n47
——, (1997b) 'Lucrèce et les Épicuriens de Campanie', in Algra, K. A.,
    Koenen, M. and Schrijvers, P., eds. (1997) *Lucretius and his Intellectual
    Background, Verh. KNAW Afd. Lett.* N. R. 172 (Amsterdam) 35-48   58n190
Dorotheus of Sidon see Stegemann (1939), Pingree (1976), Pingree (1978)
Dörrie, H. (1970) 'Der König. Ein platonisches Schlüsselwort, von Plotin mit
    neuem Sinn erfüllt', *Rev. int. philos.* 24, 217-35, repr. in Dörrie, H. (1976)
    Platonica minora (Cologne) 390-405   107n364

Dörrie, H. and Baltes, M. (1990) *Der Platonismus in der Antike. Grundlagen - System - Entwicklung* Bd. 2: *Der hellenistische Rahmen des kaiserzeitlichen Platonismus*, texts with German transl. & comm. (Stuttgart/Bad Cannstatt) 13n40

——, (1993) *Der Platonismus in der Antike. Grundlagen - System - Entwicklung* Bd. 3: *Der Platonismus im 2. und 3. Jahrhundert nach Christus*, texts with German transl. & comm. (Stuttgart/Bad Cannstatt) 82n287

Drachmann, A. G. (1948) *Ktesibios, Philon and Heron: A Study in Ancient Pneumatics*, Act. hist. sc. nat. & med. 4 (Copenhagen)   50n168

——, (1972) 'Hero of Alexandria', in Gillispie (1970-90) 6.310-4   49n161

Düring, I., ed. (1930) *Die Harmonielehre des Klaudios Ptolemaios*, Göteb. Högsk. Årsskr. 36.1 (Göteborg)   129cn225

——, (1961) *Aristotle's* Protrepticus. *An Attempt at Reconstruction*, Stud. Gr. et Lat. Gothoburgensia 12 (Stockholm)

Edelstein, L. and Kidd, I., eds. (1972), *Posidonius* vol. I: *The Fragments*, Cambr. Class. Texts & Comm. 13 (Cambridge, rev. ed. 1988)

Elias see Busse (1900)

Endress, G. (1987) 'Die wissenschaftliche Literatur', in *Grundriß der arabischen Philologie* Bd. 2, Fischer, W., ed., *Literaturwissenschaft* (Wiesbaden) 400-506   129cn260

Epicurus see Arrighetti (1973), Leone (1994)

Erotianus see Nachmanson (1918)

Etienne, E. and Roels, J. (1986) 'Deux aspects particuliers du problème des moyennes dans Pappus d'Alexandrie', *Revue des Questions Scientifiques* 157, 179-98   117n388

Euclid see Heiberg (1882), Heiberg (1883-8) Besthorn and Heiberg (1893-1932), Heiberg (1895), Menge (1896a), Menge (1916), Heath (1926), Tummers (1994)

Eudemus of Rhodos see Wehrli (1969)

Eusebius see Bardy (1952-8), Mras (1956)

Eutocius of Ascalon see Heiberg (1893), Heiberg (1915), Bulmer-Thomas (1971), Mugler (1972), Decorps-Foulquier (1997)

Fecht, R., ed. (1927) *Theodosii De habitationibus liber, De diebus et noctibis libri duo*, Abh. Göttingen phil.-hist. Kl., N. F. 19.4 (Berlin)

Ferrari, F. (1995) *Dio, idee e materia. La struttura del cosmo in Plutarcho di Cheronea*, Strum. per la Ric. Plut. 3 (Napels)   105n358

Festa, N., ed. (1891) *Iamblichus. De communi mathematica scientia liber*, Bibl. Teubn. (Leipzig, repr. with corr. and add. by Klein, U., Stuttgart 1975)

Festugière, A.-J., transl. (1967) *Proclus. Commentaire sur le Timée*, withnotes, t. 3–*livre III* (Paris)

Fihrist see Suter (1892), Dodge (1970)

Firmicus Maternus see Kroll, Skutsch and Ziegler (1913), Monat (1992-8)

Fowler, D. H. (1987) *The Mathematics of Plato's Academy: A New Reconstruction* (Oxford)   25n85

Fraser, P. M. (1972) *Ptolemaic Alexandria* 3 vols. (Oxford)   41n139   44n146   48n159 58n188

Frede, M. (1983) 'Titel, Einheit und Echtheit der aristotelischen Kategorienschrift', in Moraux and Wiesner (1983) 1-29, transl. 'The title, unity and authenticity of the Aristotelian *Categories*' in Frede (1987a) 11-28   56n185

——, (1987a) *Essays in Ancient Philosophy* (Oxford)

——, (1987b) 'Numenius', in Haase (1987b) 1034-75   101n340 106n362

Fredouille, J.-C., Goulet-Cazé, M.-O., Hoffmann, Ph. and Petitmengin, P., eds. (1997) *Titres et articulations du texte dans les oeuvres antiques*. Act. Coll. Chantilly 13-15 déc. 1994, Coll. Ét. August., Sér. Antiquité 152 (Paris)

Friderici, R. (1911) *De librorum antiquorum capitum divisione atque summariis. Accedit de Catonis De agricultura libro dissertatio* (diss. Marburg)   37n125

Friedlein, G. (1873) *Procli Diadochi In primum Euclidis Elementorum librum commentarii*, Bibl. Teubn. (Leipzig, repr. Hildesheim 1967, 1992)

Fuhrmann, M. (1960) *Das systematische Lehrbuch. Ein Beitrag zur Geschichte der Wissenschaften in der Antike* (Göttingen)   22n72 82n289

Galen see Kühn (1821-33), Von Müller (1891), Helmreich (1893), De Lacy (1978-84), Toomer (1985)

Gallo, I. (1980) *Frammenti biografici da papiri* vol. 2: *La biografia dei filosofi* (Rome)   36n124

Geminus see Tittel (1912), Aujac (1975)

Gillispie, C. C., ed. (1970-90) *Dictionary of Scientific Biography* 18 vols. (New York)

Gottschalk, H. (1965) 'Strato of Lampsacus: some texts', ed. & comm., *Proc. Leeds Philos. & Liter. Soc.* 11, 95-182   50n168 51n171

Grant, E. (1971) 'Henricus Aristippus, William of Moerbeke and two alleged medieval translations of Hero's Pneumatica', *Speculum* 46, 656-69   31n113

Grynaeus, S. and Camerarius, J., eds. (1538) *Claudii Ptolemaei Magnae Constructionis, idest Perfectae caelestium motuum pertractationis lib. XIII. Theonis Alexandrini in eosdem Commentariorum lib. XI* (Basle, *non vidi*)

Gudeman, A. (1927) 'Λύσεις', *RE* XIII (Stuttgart) 2511-29   126cn77

Gundel, W. and Gundel, H. G. (1966) *Astrologumena. Die astrologische Literatur in der Antike und ihre Geschichte*, Sudhoffs Arch. Beih. 6 (Wiesbaden)   81n283 97

Haase, W. (1982) *Untersuchungen zu Nikomachos von Gerasa* (diss. Tübingen) [includes *specimen editionis* of parts of Philoponus' Commentary on Nicomachus *Ar.*, pp. 399-447]   19n58 86n306 87n309 88n311 129cn225

———, (ed.) (1987a) *Aufstieg und Niedergang der römischen Welt*, Teil II: *Prinzipat*, Bd. 36.1: *Philosophie (Historische Einleitung; Platonismus)* (Berlin/ New York)

———, (ed.) (1987b) *Aufstieg und Niedergang der römischen Welt*, Teil II: *Prinzipat*, Bd. 36.2: *Philosophie (Platonismus [Forts.]; Aristotelismus)* (Berlin/ New York)

Hadot, I. (1978) *Le problème du Néoplatonisme alexandrin. Hiéroclès et Simplicius*, Ét. Augustin. (Paris)   107n363

———, (1984) *Arts libéraux et philosophie dans la pensée antique*, Étud. August. (Paris)   3n9 66n226 82n287 82n294 93 93n325 93n326 94 94n331

———, (1990a) *Simplicius. Commentaire sur les Catégories* fasc. 1: *Introduction, première partie (p. 1-9, 3 Kalbfleisch)*, trad. de Hoffmann, Ph. (avec la collab. de I. et P. Hadot), comm. et notes p. I. Hadot, Philos. Ant. 50 (Leiden/New York/Cologne)   1n1

———, (1990b) 'Le démiurge comme principe dérivé dans le système ontologique d'Hiéroclès', *Rev. Ét. Gr.* 103, 241-62   107n363

———, (1993) 'À propos de la place ontologique du démiurge dans le système philosophique d'Hiéroclès le Néoplatonicien. Dernière réponse à M. Ajoulat', *Rev. Ét. Gr.* 106, 430-59   107n363

Halleius, E., ed. (1706) *Apollonii Pergaei De Sectione Rationis libri duo* (Oxford)   10n29

———, ed. (1710) *Apollonii Pergaei Conicorum libri tres posteriores (sc. V^{tus} VI^{tus} & VII^{mus}) ex arabica sermone in latinum conversis, cum Pappi Alexandrini lemmatis* = Pt. 2 of *Apollonii Pergaei Conicorum libri octo et Sereni Antissensis De sectione cylindri & coni libri duo*, with Arabic text and independent pagination + Halley's reconstruction of book VIII (Oxford, repr. Osnabrück 1984) [I have only seen the original ed.]   36n125

Hasnawi, A., Elamrani-Jamal, A. and Aouad, M., eds. (1997) *Perspectives arabes et médiévales sur la tradition scientifique et philosophique grecque*, Orient. Lov. Anal. 79 (Louvain/Paris)

Hayduck, M., ed. (1888) *Asclepii in Aristotelis metaphysicorum libros A-Z commentaria*, CAG 6,2 (Berlin)

——, ed. (1891) *Alexandri Aphrodisiensis in Aristotelis Metaphysica commentaria*, CAG 1 (Berlin)

Heath, T. L. (1896) *Apollonius of Perga. Treatise on Conic Sections Edited in Modern Notation, with Introductions Including an Essay on the Earlier History of the Subject* (Cambridge)   36n125

——, (1913) *Aristarchus of Samos, the Ancient Copernicus. A History of Greek Astronomy together with Aristarchus's Treatise On the Sizes and Distances of the Moon*, new Greek text with English transl. & notes (Oxford, repr. 1959, 1981, 1997)   16n47

——, (1921) *A History of Greek Mathematics* 2 vols. (Oxford, repr. New York 1981)   6n15   12n36   36n122   47n155   49n162   49n164   55n182   58n188   62n205   62n206   86n304   118   118n393

——, transl. (1926) *The Thirteen Books of Euclid's Elements* 3 vols., with comm., 2nd. ed. (Cambridge, repr. New York 1956)   23n76   23n78   24n83   25n87   26n92   30n111   126cn89

Heiberg, J. L. (1880) 'Philologische Studien zu griechischen Mathematikern. 1. Ueber Eutokios', *Jahrbb. class. Philol.* Suppl. 11, 357-84   7n18   7n19   10n27   40n133

——, (1882) *Literargeschichtliche Studien über Euklid* (Leipzig)   25n87   27n94   34n120   58n188   58n189   58n190   64n218

——, ed. (1883-8) *Euclidis Elementa = Euclidis Opera omnia* ed. Heiberg, J. L. and Menge, H. vols. 1-5, Bibl. Teubn. (Leipzig), post Heiberg ed. Stamatis, E. S. (1977) Bibl. Teubn. (Leipzig)   125cn 77

——, ed. (1888a) *Scholia in Euclidis Elementa = Euclidis Opera omnia* ed. Heiberg, J. L. and Menge, H., vol. 5, 71-738, Bibl. Teubn. (Leipzig), post Heiberg ed. Stamatis, E. S. (1977) *Euclidis Elementa* vol. 5, pars 1, 39-343, *Scholia in libros I - V*; pars 2, *Scholia in libros VI - XIII*, Bibl. Teubn. (Leipzig)   25n87

——, (1888b) *Om Scholierne til Euklids Elementer*, avec un résumé en français, Vidensk. Sellsk. Skr., 6. Række, hist. og philos. Afd. II. 3 (Copenhagen)   26n92   27n94   28n104

——, ed. (1891-3) *Apollonii Pergaei quae graece extant cum commentariis antiquis* 2 vols., with Latin transl., Bibl. Teubn. (Leipzig, repr. Stuttgart 1974)   36n122   62

——, ed. (1893) *Eutocii commentaria in Conica*, in Heiberg (1891-3) 2.168-361

——, ed. (1894) *Simplicii in Aristotelis De caelo commentaria*, CAG 7 (Berlin)

——, ed. (1895) *Euclidis Optica, Opticorum recensio Theonis, Catoptrica, cum scholiis antiquis = Euclidis Opera omnia* ed. Heiberg, J. L. and Menge, H. vol. 7, with Latin transl., Bibl. Teubn. (Leipzig)   58   58n188

——, ed. (1896) *Sereni Opuscula* [i.e. *De sectione cylindri* and *De sectione coni*], with Latin transl., Bibl. Teubn. (Leipzig)

——, ed. (1898-1903) *Claudii Ptolemaei Opera quae exstant omnia* vol. 1.1-2, *Syntaxis Mathematica*, Bibl. Teubn. (Leipzig)   71

——, (1903) 'Paralipomena zu Euklid', *Hermes* 38, 46-74, 161-201, 321-56   23n78   25n87   27   27n96   27n97   28n104   126cn77   126cn89

——, ed. (1907) *Claudii Ptolemaei Opera quae exstant omnia* vol. 2, *Opera astronomica minora: Phaseis, Hypotheseis, Inscriptio Canobi, Procheiron kanonon diataxis, Analemma, Planisphaerium*, Bibl. Teubn. (Leipzig)

——, ed. (1910-5) *Archimedis Opera omnia cum commentariis Eutocii* 3 vols. 2nd ed., with Latin transl., Bibl. Teubn. (Leipzig, repr. with corr. by Stamatis, E. S., Stuttgart 1972)   48n159

Heiberg, J. L., ed. (1912-4) *Heronis Alexandrini Opera quae supersunt omnia* vol. 4, *Heronis Definitiones cum variis collectionibus Heronis quae feruntur Geometrica*; vol. 5, *Heronis quae feruntur Stereometrica et De mensuris*, both with German transl., Bibl. Teubn. (Leipzig, repr. Stuttgart 1976)   49n164 55

Heiberg, J. L. and Zeuthen, H. G., eds. (1912-5) *P. Tannery. Mémoires scientifiques. Sciences exactes dans l'antiquité* 3 vols. (Toulouse/Paris)

Heiberg, J. L., ed. (1914) *Theodosius Tripolites Sphaerica*, Abh. Göttingen phil.-hist. Kl., N. F. 19.2 (Berlin)

——, ed. (1915) *Eutocii Commentarii in libros De sphaera et cylindro, Commentarius in Dimensionem circuli, Commentarius in libros De planorum aequilibriis* = Heiberg (1910-15) vol. 3

——, (1925) *Geschichte der Mathematik und Naturwissenschaften im Altertum.* Handb. Altertumswiss. V.1.2 (Munich, repr. 1960)   25n87 49n162 49n166

Helmreich, G., ed. (1893) *Claudii Galeni Pergameni Scripta minora* III, Bibl. Teubn. (Leipzig, repr. Amsterdam 1967)

Hephaestion of Thebes see Pingree (1973-4)

Heraclitus see Marcovich (1978)

Heron of Alexandria see Schmidt (1899), Nix and Schmidt (1900), Schoene (1903), Heiberg (1912-4), Diels and Schramm (1918), Drachmann (1972), Tummers (1994)

Hierocles see Köhler (1974), Hadot (1978), Hadot (1990b), Hadot (1993)

Hiller, E., ed. (1878) *Theonis Smyrnaei philosophi Platonici expositio rerum mathematicarum ad legendum Platonem utilium*, Bibl. Teubn. (Leipzig)

Hintikka, J. and Remes, U. (1974) *The Method of Analysis: Its Geometrical Origin and Its General Significance*, Synthese Libr. 75 = Boston Stud. Philos. Sc. 25 (Dordrecht)   80n278 123cn26

Hoche, R., ed. (1866) *Nicomachi Geraseni Pythagoraei Introductionis arithmeticae libri ii*, Bibl. Teubn. (Leipzig)

Hogendijk, J, P. (1986) 'Arabic traces of lost works of Apollonius', *Arch. Hist. Exact Sc.* 35, 187-253   10n29

Huebner, W., ed. (1998) *Claudii Ptolemaei Opera quae exstant omnia* vol. 3.1, *Apotelesmatica* post Boll - Boer ed., Bibl. Teubn. (Stuttgart/Leipzig) [*non vidi*, because not yet published]

Huffman, C. A. (1993) *Philolaus of Croton: Pythagorean and Presocratic. A Commentary on the Fragments and Testimonia with Interpretive Essays* (Cambridge) 131cn357

Hultsch, F., ed. (1876-8) *Pappi Alexandrini Collectionis quae supersunt* 3 vols., with introd. & Latin transl. (Berlin, repr. Amsterdam 1965)   17n50, 79n274

Hijmans Jr., B. L. (1987) 'Apuleius, Philosophus Platonicus', in Haase (1987a) 395-475   106n360

Iamblichus see Festa (1891), Pistelli (1888), Pistelli (1894), De Falco (1922), Deubner (1937), Des Places (1966), Dillon (1973)

Ingenkamp, H. G. (1971) 'Zur stoischen Lehre vom Sehen', *Rhein. Mus.* 114, 240-6   128cn192

Irigoin, J. (1997) *Tradition et critique des textes grecs* (Paris)   15n45

Isnardi Parente, M., ed. (1981) *Senocrate · Ermodoro. Frammenti*, with introd., Italian transl. & comm., La Scuola di Platone 3 (Naples)

Janáček, K. (1992) *Indice delle Vite dei filosofi di Diogene Laerzio*, Acc. <<La Colombaria>> Studi 123 (Florence)   9n25

Jones, Alex. (1986a) *Pappus of Alexandria. Book 7 of the Collection*, Pt. 1. *Introd., Text & Transl.*; Pt. 2, *Comm., Index & Figures*, Stud. Hist. Mathem. & Phys. Sc. 8 (New York/Berlin/Heidelberg/Tokyo)   6n15 6n16 7n18 7n19

9n25 10n27 10n29 10n30 11n32 11n33 11n34 12n35 13n39 16n47
20n66 24n83 32 32n114 34n120 39n130 123cn26

Jones, Alex. (1986b) 'Willem of Moerbeke, the papal Greek manuscripts and the Collection of Pappus of Alexandria in Vat. gr. 218', *Scriptorium* (40) 16-31   31 31n113 32n114

Junge, G. and Thomson, W. (1930) *The Commentary of Pappus on Book X of Euclid's Elements*, Arabic text & English transl. by Thomson, introd. remarks, notes & glossary of techn. terms by Junge and Thomson, Harvard Semitic Series 8 (Cambridge MA)   26n92 27n98 28n104

Kidd, D. (1997) *Aratus Phaenomena*, ed. with introd., transl & comm., Cambr. Class. Texts & Comm. 34 (Cambridge)   38n128

Kidd, I. (1988) *Posidonius* Vol. II: *The Commentary*, (1) *Testimonia and Fragments 1-149;* (2) *Fragments 150-203*, Cambr. Class. Texts & Comm. 14A and B (Cambridge)   23n76 72n251 124cn56

Knorr, W. R. (1986) *The Ancient Tradition of Geometric Problems* (Boston/Basle/ Stuttgart)   9n26 10n28 13n39 63n209 64n219 99n366 123cn26

——, (1987) see Dijksterhuis (1956)   48n158

——, (1989) *Textual Studies in Ancient and Medieval Geometry* (Boston/Basle/ Berlin)   6n16 7n19 8n24 13 13n39 16n47 17 17n49 17n51 18 40n133 42n141 43n143 46n151 58n188 76n261 79 79n272 80 80n282

Köhler, F. W., ed. (1974) *Hieroclis in aureum Pythagoreorum carmen commentarius*, Bibl. Teubn. (Stuttgart)

Kraus, P., and Walzer, R., eds. (1951) *Galeni Compendium Timaei Platonis aliorumque dialogorum synopsis quae extant fragmenta*, Corpus Platonicum Medii Aevi, Plato Arabus 1 (London)   131cn357

Kroll, W., ed. (1899-1901) *Procli Diadochi in Platonis Rem publicam commentarii* 2 vols., Bibl. Teubn. (Berlin, repr. Amsterdam 1965)

Kroll, W., Skutsch, F. and Ziegler, K., eds. (1913) *Ivlii Firmici Materni Matheseos libri viii* 2 vols. (Leipzig, repr. with add. by Ziegler, K., Stuttgart 1968)

Kühn, C, G., ed. (1821-33) *Claudii Galeni opera omnia* 20 vols (Leipzig, repr. with epilogue & bibliogr. notes by Schubring, K., Hildesheim 1964-5)

Lachenaud, G., ed. (1993) *Plutarque. Œuvres morales* t. 12.2, *Opinions des philosophes*, with introd., French transl. & notes, Coll. Budé (Paris)

Laks, A. and Most., G., eds. (1993) *Théophraste. Métaphysique*, with introd., French transl. & notes, Coll. Budé (Paris)   56n185

Lejeune, A. (1947) 'Les lois de la réflexion dans l'Optique de Ptolémée', *Ant. Class.* 15 [1946], 241-56   59n194

——, (1948) *Euclide et Ptolémée, deux stades de l'optique geometrique grecque* (Louvain)   59n194

——, ed. (1956) *L'Optique de Claude Ptolémée dans la version latine d'après l'arabe de l'émir Eugène de Sicile*, Univ. Louvain, Rec. trav. hist. et philol. 4.8 (Louvain)   59n194

Leone, G. (1984) 'Epicuro, *Della Natura* XIV', *Cronache Ercolanesi* 14, 17-107   106n361

Lindberg, D. C. (1976) *Theories of Vision from al-Kindi to Kepler* (Chicago)   59n191

Lloyd, G. E. R. (1973) *Greek Science after Aristotle* (London)   66n223 127cn108

——, (1987) *The Revolutions of Wisdom: Studies in the Claims and Practice of Ancient Greek Science* (Berkely/Los Angeles/London, repr. 1989)   1n2

Lorenz, K. (1931) *Untersuchungen zum Geschichtswerk des Polybios* (Stuttgart)   69n238

Loria, G. (²1914) *Le scienze esatte nell' antica Grecia*, Manuali Hoepli (Milan)   15n45

Maass, E., ed. (1898) *Commentariorum in Aratum reliquiae* (Berlin, repr. 1958) 38n128

Macierowski, E. M.. transl. (1987) *Apollonius of Perga. On Cutting Off A Ratio. An Attempt to Recover the Original Argumentation through a Critical Translation of the Two Extant Medieval Arabic Manuscripts*, ed. by Schmidt, R. H. (Fairfield, Connecticut)   10n29

Mäenpää, P. (1993) *The Art of Analysis. Logic and the History of Problem Solving* (diss. Helsinki)   123cn26

——, (1997) 'From backward configuration to configurational analysis', in Otte and Panza (1997) 201-26   123cn26

Mahoney, M. M. (1972) 'Hero of Alexandria: Mathematics', in Gillispie (1970-90) 6.314-5   55n182

Manitius, C., ed. (1909) *Procli Diadochi Hypotyposis astronomicarum positionum una cum scholiis antiquis*, with German transl. (Leipzig, repr. Stuttgart 1974) 129cn260

Mansfeld, J. (1982) 'Midden-Platonisten', in *Grote Winkler Prins*, 8th entirely rev. ed. vol. 15 (Amsterdam/Brussels) 355   99n337

——, (1986) 'Diogenes Laertius on Stoic philosophy', *Elenchos* 7, 295-382, repr. in *Studies in the Historiography of Greek Philosophy* (Assen/ Maastricht 1990) 343-428   128cn192

——, (1992) *Heresiography in Context: Hippolytus' Elenchos as a Source for Greek Philosophy*, Philos. Ant. 56 (Leiden/New York/Cologne)   87n307 109n367

——, (1994) *Prolegomena. Questions to be Settled Before the Study of an Author, or a Text*, Philos. Ant. 61 (Leiden/New York/Cologne)   1n1 4n10 7n17 12n37 30n110 38n128 38n129 41n136 48n160 56n185 57n187 58n190 62n201 63n210 80n280 81n286 85n303 86n306 88n313 92n324 120n401 122cn5 128cn217

——, (1995) 'Insight by hindsight: intentional unclarity in Presocratic proems', *Bulletin Institute Classical Studies* (London) 225-32   123cn11

——, (1998a) 'Pappus, mathematicus en een beetje filosoof', *Verh. KNAW Afd. Lett. N. R.* 61.6 (Amsterdam/Oxford/New York)   34n121

——, (1998b) 'Doxographical studies, *Quellenforschung*, tabular presentation and other varieties of comparativism', forthcoming in Burkert, W., Gemelli-Marciano, L., Matelli, E. and Orelli, L., eds., *Fragmentsammlungen philosophischer Texte der Antike–Le raccolte dei frammenti di filosofi antichin*, Aporemata 3 (Göttingen)   100n338

Mansfeld, J. and Runia, D. T. (1997) *Aëtiana. The Method & Intellectual Context of a Doxographer* vol. 1: *The Sources*, Philos. Ant. 73 (Leiden/New York/Cologne)   6n13 27n93 129cn225

Mansfeld, J. see Van der Horst and Mansfeld (1974)

Marcovich, M., ed. (1978) *Eraclito. Frammenti,* with introd., Italian transl. & comm., Bibl. Stud. Super. 64 (Florence)   127cn119

Marg, W., ed. (1972) *Timaeus Locrus. De natura mundi et animae, Überlieferung, Testimonia, Text und Übersetzung*, editio maior, Philos. Ant. 24 (Leiden)

Marinus of Neapolis see Menge (1896b), Michaux (1947), Tihon (1976), Oikonomides (1977), Sambursky (1985), Masullo (1985)

Masullo, R. (1985) *Marino di Neapoli. Vita di Proclo*, Coll. Speculum, Contrib. di filol. class. (Naples)

Meijering, E. P. (1996-8) *Athanasius. Die dritte Rede gegen die Arianer*, T. 1: *Kapitel 1-25, Einleitung, Übersetzung, Kommentar*; T. II: *Kapitel 26-58, Übersetzung und Kommentar*; T. III: *Kapitel 59-67, Übersetzung, Kommentar, theologiegeschichticher Ausblick* (Amsterdam)   108n366

Menge, H., ed., (1896a) *Euclidis Data*, in *Euclidis Opera omnia* ed. Heiberg, J. L. and Menge, H. vol. 6, with Latin transl., Bibl. Teubn. (Leipzig)

Menge, H., ed., (1896b) *Marini philosophi Commentarius in Euclidis Data*, in *Euclidis Opera omnia* ed. Heiberg, J. L. and Menge, H. vol. 6, 253-7, with Latin transl., Bibl. Teubn. (Leipzig) [Greek text repr. in Oikonomides (1977) 86-106]

——, ed., (1916) *Euclidis Phaenomena*, in *Euclidis Opera omnia* vol. 8 ed. Heiberg, J. L. and Menge, H., 1-112, with Latin transl., Bibl. Teubn. (Leipzig)

Merkelbach and West see Solmsen (1970)

Merlan, Ph. (1967 and later repr.) 'Greek philosophy from Plato to Plotinus', in Armstrong, A. H., ed. (1967) *The Cambridge History of Later Greek and Early Medieval Philosophy* (Cambridge)   99n337

Michaux, M. (1947) *Le commentaire de Marinus aux Data d'Euclide. Étude critique*, Univ. Louvain, Rec. Hist. & Philol. 3e sér. fasc. 25 (Louvain)   [with French transl. pp. 54-65, repr. in Oikonomides (1977) 87-107]   61n196 61n197 61n199 62n202 63n211 64n216

Mogenet, J. (1950) *Autolycus de Pitane. Histoire du texte*, with éd. crit. de *La sphère en mouvement* et des *Levers et couchers*, Univ. Louvain, Rec. Hist. & Philol. 3e série fasc. 37 (Louvain)   15n44 16n47

——, (1956) *L'introduction à l'Almageste*, Mém. Acad. Belg., Cl. Lettres, 2e série (Brussels)   2n6 17n50 78n267 78n269 79n271 79n272 79n273 79n276 80n279

Mogenet, J. and Tihon, A., eds. (1985) *Le "Grand Commentaire" de Théon d'Alexandrie aux Tables Faciles de Ptolémée, Livre I*, with introd., French transl. & comm., Studi e Testi 315 (Città del Vaticano)   78n268

Monat, P., ed. (1992-7) *Firmicus Maternus. Mathesis* t. 1: *Livres I-II*, t. 2: *Livres III, IV et V*, t. 3: *Livres VI-VIII*, with introd., French transl. & notes, Coll. Budé (Paris)

Moraux, P. (1963) 'Quinta essentia', *RE* XXIV   (Stuttgart) 1171-1263   131cn357

Moraux, P. and Wiesner, J., eds. (1983) *Zweifelhaftes im Corpus Aristotelicum*, Akt. 9. Symp. Arist., Peripatoi 14 (Berlin/New York)

Moreschini, C., ed. (1991) *Apuleius. De philosophia libri*, Bibl. Teubn. (Stuttgart/Leipzig)

Morrow, G. R., transl. (1970) *Proclus. A Commentary on the First Book of Euclid's Elements*, with introd. & notes (Princeton, repr. with new foreword by Mueller, I., 1992)

Mras, K., ed. (1956) *Eusebius Werke* Bd. 8.1-2: *Die Praeparatio Evangelica*, GCS 43.1 & 43.2 (repr. Berlin 1982-3)

Mueller, I. (1987) 'Mathematics and philosophy in Proclus' commentary on book I of Euclid's Elements', in Pépin and Saffrey (1987) 305-18   24n82

——, (1992) see Morrow (1970)   24n80

Mugler, Ch., ed. (1972) *Archimède* t. 4: *Commentaires d'Eutocius et Fragments*, with French transl., Coll. Budé (Paris)   103n349

Mutschmann, H. (1911) 'Inhaltsangabe und Kapitelüberschrift im antiken Buch', *Hermes* 46, 93-107   69n238 129cn225

Mynors, R., ed. (1937) *Cassiodorus. Institutiones divinarum et humanarum litterarum* (Oxford)

Nachmanson, E., ed. (1918) *Erotianus. Vocum hippocraticorum collectio cum fragmentis*, Coll. script. vet. upsaliensis (Göteborg)

Nauck, A., ed. (1886) *Porphyrius. Vita Pythagorae*, in *Porphyrii philosophi Platonici opuscula selecta*, Bibl. Teubn. (Leipzig, repr. Hildesheim 1963)

Neugebauer, O. (1938) 'Über eine Methode zur Distanzbestimmung Alexandria–Rom bei Heron', *Kgl. Danske Vidensk. Selsk., Hist.-filol. Medd.* 26.2 (Copenhagen)   49n161

Neugebauer, O. (1975) *A History of Ancient Mathematical Astronomy* 3 vols., Stud. Hist. Mathem. & Phys. Sc. 1 (Berlin/Heidelberg/New York)   14n41

144   BIBLIOGRAPHY

14n42 16n47 18n54 20n64 23n79 49n161 54n177 58n188 59n194
71n247 76n263 127cn192 129cn260
Nicomachus of Gerasa see Hoche (1866), Pistelli (1894), Von Jan (1895-8),
    D'Ooge (1926), Tarán (1969), Tarán (1974), Bertier (1978), Haase (1982),
    Zanoncelli (1990)
Nix, L., ed. (1889) *Das fünfte Buch der Conica des Apollonius von Perga in der
    arabischen Uebersetzung des Thabit ibn Corrah*, with introd. and German
    transl. (Leipzig)
Nix, L. and Schmidt, W., eds. (1900) *Heronis Alexandrini Opera quae supersunt
    omnia* vol. 2, *Mechanica et Catoptrica*, with German transl., Bibl. Teubn.
    (Leipzig, repr. Stutgart 1976)   52n173
Nobbe, C. F. A., ed. (1843-5) *Claudii Ptolemaei Geographia* (repr. Hildesheim
    1966, with introd. by Diller, A.)
Numenius see Des Places (1973), Frede (1987b)
O'Brien, D. (1992) 'Origène et Plotin sur le roi de l'univers', in *ΣΟΦΙΗΣ
    ΜΑΙΗΤΟΡΕΣ*, << *Chercheurs de sagesse* >>, Hommage à Jean Pépin, Coll. Ét.
    August., Sér. Ant. 131 (Paris) 317-42   101n340 107n363 107n364
O'Meara, D. J. (1989 and later repr.) *Pythagoras Revived. Mathematics and
    Philosophy in Late Antiquity* (Oxford)   4n9   24n82   82n287   83n295
    86n304 89n318 109n367 119n399 128cn225 130cn308
Oikonomides, A. N., ed. (1977) *Marinus of Neapolis. The Extant Works or The Life
    of Proclus and the Commentary on the* Dedomena *of Euclid*, Greek text with
    (English and French) transl., Testimonia de vita Marini, introd. &
    bibliogr. (Chicago)
Oliver, R. P. (1951) 'The first Medicean ms of Tacitus and the titulature of
    ancient books', *Trans. Proc. Amer. Phil. Ass.* 82, 232-61   18n55
Olympiodorus see Stüve (1900), Busse (1902), Westerink (1956), Westerink
    (1970)
Otte, M. and Panza, M., eds. (1997) *Analysis and Synthesis in Mathematics.
    History and Philosophy*, Boston Stud. Hist. Sc. 196 (Dordrecht/Boston/
    London)
Ouzounian, A. (1994) 'David l'invincible', in Goulet, R., ed. (1994) *Dictionnaire
    des philosophes antiques* t. 2, *Babélyca d'Argos à Dyscolius* (Paris) 614-5
    67n228
Panza, M. (1997) 'Classical sources for the concepts of analysis and synthesis',
    in Otte and Panza (1997) 365-414   9n25
Pappus of Alexandria see Woepcke (1856), Hultsch (1876-8), Suter (1922),
    Junge and Thomson (1930), Rome (1931), Ver Eecke (1933), Treweek
    (1950), Bulmer-Thomas (1974), Hintikka and Remes (1974) Jones
    (1986a), Jones (1986b), Mäenpää (1993)
Patillon, M. and Bolognesi, G., eds. (1997) *Aelius Théon. Progymnasmata*, with
    introd., French transl. & notes, Coll. Budé (Paris)   122cn5
Pease, A. S., ed. (1955-8) *M. Tvlli Ciceronis De Natura Deorum*, with introd. &
    copious notes, 2 vols. (Cambridge MA, repr. New York 1979)   109n368
Petitmingin, P. (1997) 'Capitula païens et chrétiens', in Fredouille *& al.*
    (1997) 491-507   129cn225
Philoponus see Hayduck (1897), Rabe (1899), Wallies (1909), Segonds (1981),
    Haase (1982)
Pingree, D. (1968) rev. De Falco, V., Krause, M. and Neugebauer, O.,
    Hypsikles. Die Aufgangszeiten der Gestirne, *Gnomon* 40, 13-7   16n47
——, ed. (1973-4) *Hephaestionis Thebani Apotelesmatica* 2 vols., Bibl. Teubn.
    (Leipzig)   74n258 96
——, ed. (1976) *Dorothei Sidonii Carmen Astrologicum. Interpretationem arabicam in
    linguam anglicam versam una cum Dorothei fragmentis et graecis et latinis*, Bibl.
    Teubn. (Leipzig)   97

Pingree, D. (1978) 'Dorotheus of Sidon', in Gillispie (1970-90) 15.125   97
——, (1997) 'Masha allah: Greek, Pahlavi, Arabic and Latin astrology', in Hasnawi & al. (1997) 123-36   97
Pistelli, H., ed. (1888) Iamblichus. Protrepticus, Bibl. Teubn. (repr. Stuttgart 1957)
——, ed. (1894) Iamblichi in Nicomachi Arithemeticam introductionem liber, Bibl. Teubn. (Leipzig, 2nd ed. with corr. & add. by Klein, J., Stuttgart 1975) 87n308
Plutarch see Cherniss (1976), Ferrari (1995)
Porphyry see Nauck (1886), Boer and Weinstock (1940), Smith (1993)
Posidonius see Edelstein and Kidd (1972), Kidd (1988)
Proclus see Allatius (1731), Friedlein (1873), Van Pesch (1900), Kroll (1899-1901), Diehl (1903-6), Manitius (1909), Festugière (1967), Saffrey and Westerink (1968-98), Morrow (1970), Steel (1982-5), Siorvanes (1996)
Ptolemaeus Gnosticus see Quispel (1966)
Ptolemy see Allatius (1731), Nobbe (1843-5), Boll (1894), Heiberg (1898-1903), Heiberg (1907), Düring (1930), Boer and Weinstock (1940), Boll and Boer (1940), Robbins (1940), Lejeune (1956), Toomer (1975), Toomer (1984), Smith (1988), Simon (1997), Huebner (1998)
Quispel, G., ed. (^{2}1966) Ptolémée. Lettre à Flora, SC 24bis (Paris)
Rabe, H., ed. (1899) Ioannes Philoponus. De aeternitate mundi contra Proclum, Bibl. Teubn. (Leipzig, repr. Hildesheim 1963)
——, ed. (1931) Prolegomenon Sylloge, Rhet. gr. 14, Bibl. Teubn. (Leipzig, repr. Stuttgart/Leipzig 1995)
Remes, U. see Hintikka and Remes (1974)
Richard, M. (1950) 'ΑΠΟ ΦΩΝΗΣ', Byzantion 20, 191-222, repr. in Opera minora vol. 3 (Turnhout/Louvain 1977) No. 60   58n190
Robbins, F. E., ed. (1940 and later repr.) Ptolemy Tetrabiblos, with English transl., LCL 350 pt. 2 (London/Cambridge MA)   71n244 75n259
Rome, A., ed. (1931) Commentaires de Pappus et de Théon d'Alexandrie sur l'Almageste t. 1: Pappus. Commentaire sur les livres 5 et 6 de l'Almageste, texte établi et annoté, Studi e Testi 54 (Rome, repr. Città del Vaticano 1967)   76n262 129cn225
——, ed. (1936) Commentaires de Pappus et de Théon d'Alexandrie sur l'Almageste t. 2: Théon d'Alexandrie. Commentaire sur les livres 1 et 2 de l'Almageste, texte établi et annoté, Studi e Testi 72 (Città del Vaticano)
——, ed. (1943) Commentaires de Pappus et de Théon d'Alexandrie sur l'Almageste t. 3: Théon d'Alexandrie. Commentaire sur les livres 3 et 4 de l'Almageste, texte établi et annoté, Studi e Testi 106 (Città del Vaticano)
——, (1953) Sur l'authenticité du 5e livre du Commentaire de Théon d'Alexandrie sur l'Almageste, Mém. Acad. Belg., Cl. Lettr., 2e série t. 39 (Brussels) 76n261
Ross, W. D. (1949 and later repr.) Aristotle's Prior and Posterior Analytics, rev. text with introd, & comm. (Oxford)   83n295
Runia, D. T. (1986) Philo of Alexandria and the Timaeus of Plato, Philos. Ant. 44 (Leiden/New York/Cologne)   105n358
——, (1997) 'The literary and philosophical status of Timaeus' prooemium', in Calvo, T. and Brisson, L., Interpreting the Timaeus–Critias, Proceed. 4th Symp. Plat., Intern. Plato Stud. 9 (Graz) 101-18   123cn11
——, see Mansfeld and Runia (1997)
Sachs, E. (1917) Die fünf platonischen Körper. Zur Geschichte der Mathematik und der Elementenlehre Platons und der Pythagoreer (Berlin, repr. New York 1976) 27n98 27n99 33n119
Saffrey, H. D. and Westerink, L. G., eds. (1968-98) Proclus. Théologie platonicienne with introd., French transl., notes & indices, 6 vols., Coll. Budé (Paris)   129cn225

Sambursky, S. (1985) *Proklos, Präsident der platonischen Akademie, und sein Nachf-
olger, der Samaritaner Marinos* (Berlin/Heidelberg/New York/Tokyo)
61n199
Schissel von Fleschenberg, O. (1930) 'Marinos (Neuplat.)', *RE* XIV (Stuttgart)
1759-67   2n6 61n196 61n197 61n199 62n202 72n250
Schmidt, W., ed. (1899) *Heronis Alexandrini Opera quae supersunt omnia* vol. 1:
*Pneumatica et Automata. Accedunt fragmentum De horoscopiis aquariis, Philonis
De ingeniis spiritualibus, Vitruvii capita quaedam ad Pneumatica pertinentia,* with
comm. & German transl., Bibl. Teubn. (Leipzig, repr. Stuttgart 1976)
Schöne, H., ed. (1903) *Heronis Alexandrini Opera quae supersunt omnia* vol. 3:
Rationes dimetiendi et Commentatio dioptrica, with German transl., Bibl.
Teubn. (Leipzig, repr. Stuttgart 1976)
Schrenk, L. P. (1994) 'Proof and discovery in Aristotle and the later Greek
tradition: a prolegomenon to a study of analysis and synthesis', in
Schrenk, L. P., ed., *Aristotle in Late Antiquity,* Stud. Philos. & Hist. Philos.
27 (Washington D.C.) 92-108   123cn26
Sedley, D. (1976) 'Epicurus and the mathematicians of Cyzicus', *Cronache
Ercolanesi* 6, 23-54   23n76
——, (1989) 'Philosophical allegiance in the Greco-Roman world', in Barnes,
J. and Griffin, M., eds., *Philosophia Togata. Essays on Philosophy and Roman
Society* (Oxford, rev. repr. as *Philosophia Togata* I, 1997) 97-119   58n190
Segonds, A. P. (1981) *Jean Philopon. Traité de l'astrolabe,* with introd., French
transl. & text [141-156, repr. of ed. by Hase, H. (Bonn 1839), with
corrigenda *ibid.* 299-303] (Paris)   88n313
——, (1987) 'Proclus: astronomie et philosophie', in Pépin and Saffrey (1987)
319-34   129cn260
Serenus of Antinoupolis see Heiberg (1896), Decorps-Foulquier (1992)
Sezgin, F. (1974) *Geschichte des arabischen Schrifttums* Bd. 5: *Mathematik bis ca.
430 H.* (Leiden)   25n84 26n90 125cn77
——, ed. (1986) *Heinrich Suter. Beiträge zur Geschichte der Mathematik und
Astronomie im Islam* 2 Bde. (Francfort)
Shiel, J. (1990) 'Boethius' Commentaries on Aristotle', rev. repr. in Sorabji
(1990) 349-72   13n39
Simon, G. (1988) *Le regard, l'être et l'apparence dans l'Optique de l'Antiquité*
(Paris)   59n194 127cn192 128cn192
——, (1997), 'La psychologie de la vision chez Ptolémée et Ibn al-Haytham', in
Hasnawi *& al.* (1997) 189-208   127cn192
Simplicius see Diels (1882), Heiberg (1894), Tummers (1994)
Siorvanes, L. (1996) *Proclus. Neoplatonic Philosophy and Science* (Edinburgh)
129cn260
Smith, A., ed. (1993) *Porphyrii philosophi fragmenta,* fragmenta arabica D.
Wasserstein interpretante, Bibl. Teubn. (Stuttgart/Leipzig)
Smith, A. M. (1988) 'The psychology of visual perception in Ptolemy's *Optics*',
Isis 79, 189-207   59n194
Solmsen, F., Merkelbach, R., and West, M. L. (1970) *Hesiodi Theogonia Opera
et Dies Scutum* [Solmsen] *Fragmenta selecta* [Merkelbach and West]
(Oxford)
Sorabji, R., ed. (1990) *Aristotle Transformed: The Ancient Commentators and their
Influence* (London)
Speusippus see Tarán (1981)
Steel, C., ed. (1982-5) *Proclus Diadochus. Commentaire sur le Parménide de Platon,
trad. de Guillaume de Moerbeke,* Ancient & Med. Philosophy ser. 1, 3-4, t. 1:
Livres I à IV, t. 2: *Livres V à VII et Notes marginales de Nicolas de Cues*
(Leuven)

Stegemann, V., ed. (1939) *Die Fragmente des Dorotheus von Sidon*, Quellen und Stud. zur Gesch. und Kult. des Altert. und des Mittelalt., Reihe B. 1, with comm. (Heidelberg) 97

Stüve, G., ed. (1900) *Olympiodori in Aristotelis meteora commentaria*, CAG 12.2 (Berlin)

Suter, H. (1892) 'Das Mathematiker-Verzeichnis im Fihrist des Ibn Abi Ja'kub an-Nadim. Zum ersten Mal vollständig ins Deutsche übersetzt und mit Anmerkungen versehen', *Zeitschr. Phys. Math.* 37, Suppl., 1-87, repr. in Sezgin (1986) 1.315-404 25n84 26n90 126cn77 126cn89

———, transl. (1922) 'Der Kommentar des Pappus zum X. Buche des Eukleides', *Abh. Gesch. Naturwiss. Mediz.* H. 4, 9-78 (Erlangen), repr. in Sezgin (1986) 2.550-619 24n84 25n85 27n98 32n114

Swift Riginos, A. (1976) *Platonica. The Anecdotes Concerning the Life and Writings of Plato*, Columbia Stud. Class. Trad. 3 (Leiden) 68n229

Tannery, P. (1882) 'Sur les fragments de Héron d'Alexandrie conservés par Proclus', in Heiberg and Zeuthen (1912-5) 1.156-67 125cn77

———, ed. (1895) *Anonymi prolegomena in Introductionem arithmeticam Nicomachi*, in *Diophanti Alexandrini Opera omnia cum graecis commentariis* 2.73-7, Bibl. Teubn. (Leipzig) 89n318

Tarán, L., ed. (1969) *Asclepius of Tralles, Commentary to Nicomachus' Introduction to Arithmetic*, with. introd. & notes, Trans. Am. Phil. Soc. N. S. 59.4 (Philadelphia) 88n311 88n312 89n318

———, (1974) 'Nicomachus of Gerasa', in Gillispie (1970-90) 10.112-4 82n287

———, (1981) *Speusippus of Athens. A Critical Study with a Collection of the Related Texts and Commentary*, Philos. Ant. 39 (Leiden) 131cn357

Tarrant, H. (1995) 'Introducing philosophers and philosophies', *Apeiron* 28, 141-58 [rev. Mansfeld 1994] 5

Taub, L. C. (1993) *Ptolemy's Universe. The Natural Philosophical and Ethical Foundations of Ptolemy's Astronomy* (Chicago/La Salle, repr. 1994) 66n226 71n247 71n248

Theiler, W. (1930) *Die Vorbereitung des Neuplatonismus*, Problemata 1 (Berlin, repr. 1964) 99n337

Theodosius see Heiberg (1914), Fecht (1927)

Theon of Alexandria see Grynaeus and Camerarius (1538), Rome (1936), Toomer (1976b), Rome (1953), Tihon (1978), Mogenet and Tihon (1985), Tihon (1991)

Theon of Smyrna see Hiller (1878)

Thesleff, H.., ed., (1965) *The Pythagorean Texts of the Hellenistic Period*, Acta Academiae Aboensis, Humaniora, Ser. A 30.1 (Å bo) [text of Timaeus Locrus ed. Marg, W.]

Thomas, I., ed, (1939-41) *Greek Mathemtical Works*, LCL 2 vols. (Cambridge MA/London, repr. 1957)

Thomson (1930) see Junge and Thomson (1930) 24n84 27n95 27n98 28n100 36n123

Thomson, R. W., ed. (1971) *Athanasius: Contra gentes et de incarnatione*, Oxford Early Christian Texts (Oxford)

Tihon, A., (1976) 'Notes sur l'astronomie grecque au Ve siècle de notre ère (Marinus de Naplouse–un commentaire au *Petit commentaire* de Théon)', *Janus* 63, 167-84 65n222 129cn225

———, ed. (1978) *Le "Petit Commentaire" de Théon d'Alexandrie aux Tables Faciles de Ptolémée* (Histoire du texte, édition critique, traduction), Studi e Testi 282 (Città del Vaticano) 76n261 78n264 78n270

———, ed. (1991) *Le "Grand Commentaire" de Théon d'Alexandrie aux Tables Faciles de Ptolémée, Livres II et III* (Histoire du texte, édition critique, traduction), Studi e Testi 340 (Città del Vaticano) 78n267

148   BIBLIOGRAPHY

Timaeus Locrus see Marg (1972), Baltes (1972)

Tittel, K. (1912) 'Geminos', *RE* VII (Stuttgart) 1026-50   24n81

Todd, R., ed. (1990) *Cleomedis Caelestia (METEΩPA)*, Bibl. Teubn. (Leipzig) 24n81

Toomer, G. J. (1970) 'Apollonius of Perga', in Gillispie (1970-90) 1.179-83 36n122 37n127

——, (1975) 'Ptolemy', in Gillispie (1970-90) 11.186-206   66n223

——, ed. (1976a) *Diocles on Burning Mirrors. The Arabic Translation of the Lost Greek Original*, with English transl. & comm., Sources Hist. Math. Phys. Sc. 1 (Berlin/Heidelberg/New York)   33n117

——, (1976b) 'Theon of Alexandria', in Gillispie (1970-90) 13.321-5 25n87 58n188 76n261 77n265

——, transl. (1984) *Ptolemy's Almagest*, with notes (London)   69n234 69n237 129cn225

——, (1985) 'Galen on the astronomers and astrologers', *Arch. Hist. Exact Sc.* 32, 193-206   15n46

——, (1990) *Apollonius Conics Books V to VII. The Arabic Translation of the Lost Greek Original in the Version of the Banu Musa*, Pt. 1. Introd., Text & Transl.; Pt. 2, Comm., Figures & Indexes, Sources Hist. Math. Phys. Sc. 9 (New York/ Berlin etc.)   36n122 36n125 38n128 41n133

Treweek, A. P., ed. (1950) *A Critical Edition of the Text of the Collection of Pappus of Alexandria, Books II to V* (handwritten diss. London, microfilm)

Tummers, P. M. J. E. (1984) *Albertus (Magnus)' Commentaar op Euclides' Elementen der Geometrie*, I: *Inleidende Studie en Analyse*; II: *Uitgave van boek I van Albertus (Magnus) en van Anaritius*, diss. Leiden (Nijmegen)   26n90

——, ed. (1994) *The Latin Translation of Anaritius' Commentary on Euclid's Elements of Geometry, Books I-IV*, Artistarium Suppl. 9 (Nijmegen)   26n90

Van Berchem, D. (1952) 'Poètes et grammariens. Recherches sur la tradition scolaire d'explication des auteurs', *Museum Helveticum* 9, 79-87   63n210

Van der Horst, P. W. (1996) ''A simple philosophy': Alexander of Lycopolis and Christianity, in Algra & al. (1996) 313-29   108n365

Van der Horst, P. W. and Mansfeld, J. (1974) *An Alexandrian Platonist against Dualism: Alexander of Lycopolis' treatise 'Critique of the Doctrines of Manichaeus'*, transl. with introd. & notes (Leiden), introd. repr. with same pagin. in Mansfeld, J. (1990) *Studies in the Historiography of Greek Philosophy* (Assen/ Maastricht)   108n365

Van Pesch, J. G. (1900) *De Procli fontibus. Dissertatio ad historiam matheseos graecae pertinens* (diss. Leiden)   24n80 125cn77

Van Sickle, J. (1980) 'The book-roll and some conventions of the poetic book', *Arethusa* 13, 5-42   69n238 123cn11

Vanhamel, W. (1989) 'Bibliographie de Guillaume de Moerbeke', in Brams, J. and Vanhamel, W., eds. (1989) *Guillaume de Moerbeke. Recueil d'études à l'occasion du 700ᵉ anniversaire de sa mort (1286)*, Anc. & Med. Philos. 1.7 (Leuven) 301-83   32n113 45n148 52n173 97

Ver Eecke, P., transl. (1933) *Pappus d'Alexandrie. La Collection mathématique*, with introd. & notes 2 vols. (Paris/Bruges, repr. Paris 1982)   6n15 118n392

Verrycken, K. (1994) *De vroege Philoponus. Een studie van het Alexandrijns Neoplatonisme*, Verh. Ak. Belg. Kl. Lett. 56, Nr. 153 (Brussels)   43n143

Villey, A., transl. (1985) *Alexandre de Lycopolis. Contre la doctrine de Mani*, with comm., Sources Gnostiques et Manichéennes 2 (Paris)

Vlastos, G. (1975) *Plato's Universe* (Oxford)   109n368

Von Jan, C., ed. (1895-8), *Nicomachi Geraseni Harmonica*, in *Musici scriptores graeci. Aristoteles, Euclides, Nicomachus, Bacchius, Gaudentius, Alypius et melodiarum veterum quidquid exstat*, Bibl. Teubn. (repr. Hildesheim 1972) 237-65 [also see Zanoncelli (1990)]

Von Müller, I., ed. (1891) *Claudii Galeni Pergameni Scripta minora* II, Bibl. Teubn. (Leipzig, repr. Amsterdam 1967)

Vuillemin-Diem, G. (1983) 'Anmerkungen zum Pasiklesbericht und zu Echt-heits-Zweifeln am größeren und kleineren Alpha in Handschriften und Kommentaren', in Moraux and Wiesner (1983) 157-92   124cn67

Wallies, M., ed. (1883) *Alexandri in Aristotelis Analyticorum priorum librum i commentarium*, CAG 2.1 (Berlin)

———, ed. (1891) *Alexandri Aphrodisiensis in Aristotelis Topicorum libros octo commentaria*, CAG 2.2 (Berlin)

———, ed. (1909) *Ioannis Philoponi in Aristotelis Analytica Posteriora commentaria cum Anonymo in librum ii*, CAG 13.3 (Berlin)

Waszink, J. H., ed. (1962) *Timaeus a Calcidio translatus commentarioque instructus*, Corpus Platonicum Medii Aevi, Plato Latinus 4 (London/Leiden, 2nd rev. ed. 1976)

Weber, K. O. (1962) *Origenes der Neuplatoniker. Versuch einer Interpretation*, with the testimonia, Zetemata 27 (Munich)   107n364

Wehrli, F., ed. (1969) *Eudemos von Rhodos*. Schule des Aristoteles H. 8 (Basle, 2nd ed.)   32n115 126cn89

Westerink, L. G., ed. (1956) *Olympiodorus. Commentary on the First Alcibiades of Plato* (Amsterdam, repr. 1982)

———, ed. (1970) *Olympiodori in Platonis Gorgiam commentaria*, Bibl. Teubn. (Leipzig)

———, ed. (1977) *The Greek Commentaries on Plato's Phaedo* vol. 2: *Damascius*, Verh. KNAW Afd. Lett. N. R. 93, with introd. & notes (Amsterdam/Oxford/ New York)   114n380

———, ed. (1985) *Stephanus of Athens. Commentary on Hippocrates' Aphorisms, Sections I-II*, CMG XI 1,3,1, with introd. & German transl. (Berlin)   1n1

———, (1990) 'The Alexandrian commentators and the introductions to their commentaries', in Sorabji (1990) 325-48   4n10

Whittaker, J. (1981) 'Plutarch, Platonism and Christianity', in Blumenthal, H. J. and Markus, R. A., eds., *Neoplatonism and Christian Thought: Essays in Honour of A. H. Armstrong* (London) 50-63, repr. as Study XVIII in Whittaker, J. (1981) *Studies in Platonism and Patristic Thought* (London)   107n364

———, ed. (1990) *Alcinoos. Enseignement des doctrines de Platon*, with introd., French transl. (by Louis, P.) & notes, Coll. Budé (Paris)   103n348

Woepcke, F. (1855) editio princeps of Arabic text of *Pappus in Eucl. Elem.* X (Paris, *non vidi*)   24n84

———, (1856) *Essai d'une restitution de travaux perdus d'Apollonius sur les quantités irrationelles*, Mémoires présentés par divers savants à l'Académie des Sciences de l'Institut impérial de France t. 14 (Paris) 685-720 [*non vidi*, but for Pappus see Heiberg (1891-3) 2.120-4]   24n84

Xenocrates see Isnardi Parente (1981)

Zanoncelli, L. (1990) *La manualistica musicale greca. [Euclide]. Cleonide. Nicomaco. Excerpta Nicomachi. Bacchio il Vecchio. Gaudenzio. Alipio. Excerpta Neapolitana*, Von Jan's (1895-9) texts with introd., Italian transl. & comm. (Milan)

Ziegler, K. (1934) 'Theon von Alexandreia', *RE* VA (Stuttgart) 2075-80   25n87 58n188

———, (1949) 'Pappos von Alexandreia', *RE* XVIII (Stuttgart) 1084-1106   6n15 6n16 8n23 14n43 28n102 76n264

Ziegler, K., Boer, E., Lammert, F., and Van der Waerden, B. L. (1959) 'Klaudios Ptolemaios, der Astronom und Geograph', *RE* XXIII.2 (Stuttgart) 1788-1859   66n223 96n335

Zintzen, C., ed. (1967) *Damascii Vitae Isidori reliquiae*, with notes (Hildesheim)

INDEX LOCORUM POTIORUM

The numbers refer to the location in the footnotes (abbreviated n, e.g. 47n156 means p. 47 note 156), in the complementary notes (abbreviated cn, e.g. 122cn5 means p. 122 complementary note 5), and in the text (just the page number).

154   INDEX LOCORUM

CASSIODORUS

Institutiones (Mynors)
2.7.2   17-8, 18n54

CICERO

Ad Atticum
4.16.2   123cn11

De natura deorum
2.47   109n368

De oratore
                     18
1.47   5

Lucullus
106   23n76

Orator
                     18

Timaeus
17   109n368

CLEMENT OF ALEXANDRIA

Stromata (Stählin, Früchtel and
Treu)
1.21.104.2   125cn67
6.8.64   128cn217
7.5.28   104n355

DAMASCIUS

In Phaedonem (Westerink)
123.7   73n254
516   114n380

Vitae Isidori reliquiae (Zintzen)
                     19
Fr. 109.12-5   19n57
p. 191   129cn260
p. 199   61n196,
                     61n200

DAVID

In Isagogen (Busse)
p. 80.13   128cn217
p. 95.9-10   128cn217

Prolegomena (Busse)
p. 5.6-8   67n228
p. 5.13-7   68n229
p. 26.9 ff.   130cn308
p. 44.5-6   21n71

p. 57.21-2   68n229
p. 59.12-23   68n229
p. 59.26-32   68n229

DAVID (ELIAS ?)

In Categorias (Busse)
p. 80.13   128cn127
p. 95.9-10   128cn127
p. 115.18-9   67n228

DIOCLES

On Burning Mirrors (Toomer)
prooem.   33n117

DIODORUS SICULUS

Bibliotheca historica
                     69n238
1.8.5   21n71
1.23.7   19n60
1.29.6   19n60
1.52   122cn11
1.86.3   19n60
1.98.10   122cn11
3.61.6   19n60
3.74.6   122cn11
4.85.7   122cn11
13.114.3   122cn11
14.117.9   122cn11
15.95.4   122cn11
16.95.5   122cn11
17.118.3   122cn11
18.75.3   122cn11
19.110.8   122cn11

DIOGENES LAËRTIUS

1.34   20n66
3.50   125cn67
3.65-6   13n40
3.69   112n377
3.72   112, 112n376,
                     114
5.23   125cn67
5.28   67n227
5.29   125cn67
5.50   56n185
7.41   128cn192
7.48 ff.   128cn192
7.157   128cn192
8.35   102n343
9.31-2   110n369
9.41   56

INDEX RERUM ET NOMINUM ANTIQUORUM

For *nomina antiqua* see also *index locorum potiorum*. The numbers again refer to the location in the footnotes (abbreviated n, e.g. 47n156 means p. 47 note 156), in the complementary notes (abbreviated cn, e.g. 129cn225 means p. 129 complementary note 225), and in the text (just the page number). The cross-references in the notes may also be of some help.

Ptolemy 69, 78
Theon of Alexandria 78
see also mathematics, teaching
of
• qualities of the student 5
Aristotle 4n10
Eutocius 42
Galen 123-4cn56
Heron of Alexandria 50
Nicomachus of Gerasa 85,
85n300
Pappus of Alexandria 8, 12
Philoponus 88n313
Ptolemy 68, 69, 78, 79
Theon of Alexandria 77, 78,
79
• systematic organisation see
arrangement (τάξις)
• theme (aim, contents,
authorial intention, purpose,
subject, περιοχή, πρόθεσις,
σκόπος, τέλος, ὑπόθεσις) 4
Aelius Theon 122cn5
Anon. in Nicom. 89, 90
Anon. in Ptol. 79
Apollonius of Perga 37-8,
39, 42
Archimedes 44, 45, 46, 47
Aristotle 4n10
Asclepius 88
Diodorus Siculus 69n238,
122cn11
Dionysius of Halicarnassus
122cn11
Euclid 29, 30, 32, 34, 63
Eutocius 42, 45, 47, 123cn11
Heron of Alexandria 51, 52
Marinus of Neapolis 62
Nicomachus of Gerasa 82-3,
89, 90
Pappus of Alexandria 11, 12,
14, 20, 32, 34, 39, 46, 77
Philoponus 88
Polybius 69n238
ps.Longinus 72n250
Ptolemy 68, 69, 71, 72, 74,
74n257, 75, 77, 79, 122cn11
Scholia in Euclidem 29, 30
historical note / overview
4, 13n39, 21, 29, 30, 32, 33-
4, 37, 38, 40, 41, 46, 48, 53,
54, 62, 64, 69, 86-7
see also predecessors
• title (explanation / justifica-
tion / authenticity of titles;
ἐπιγραφή) 4
Anon. in Ptol. 80
Apollonius of Perga 10, 21
Archimedes 46
Aristaeus 10

Aristarchus 14
Autolycus 14
Eratosthenes 11, 21
Euclid 10-1, 14, 30, 32, 58,
62, 63, 126cn108
Eutocius 46
Galen 123-4cn56
Marinus of Neapolis 62
Nicomachus of Gerasa 82,
88-9
Pappus of Alexandria 7, 10-
11, 21, 32
Philoponus 88-9
Ptolemy 17, 17n52, 77-8, 80
Scholia in Euclidem 30,
126cn108
Theodosius 14
Theon(?) of Alexandria 58,
77-8
see also title(s)
• to which part of mathematics a
work / sub-discipline
belongs 5
Aelius Theon 122cn5
Anon. in Ptol. 80
Euclid 29, 63
Heron of Alexandria 21-2,
50, 51, 52, 54
Marinus of Neapolis 62, 63
Nicomachus of Gerasa 9n25
Pappus of Alexandria 9, 14,
21-2
Ptolemy 9n25, 22n72, 66-7,
80
Scholia in Euclidem 29
see also mathematics, division
of
• utility (χρήσιμον; ὠφέλεια etc.)
4
Aelius Theon 122cn5
Anon. in Nicom. 89, 90
Anon. in Ptol. 80
Apollonius of Perga 37, 40
Archimedes 46, 47
Euclid 29, 33, 34, 62, 63
Eutocius 46, 47
Galen 128cn217
Heron of Alexandria 50, 51,
52, 53, 54
Marinus of Neapolis 62, 63
Nicomachus of Gerasa 82,
84, 89, 90
Pappus of Alexandria 20-1,
33, 34
Philoponus 88n313
ps.Longinus 72n250
Ptolemy 66, 68, 69, 72, 78,
80
Scholia in Euclidem 29
Theon of Alexandria 78

PHILOSOPHIA ANTIQUA

A SERIES OF STUDIES ON ANCIENT PHILOSOPHY

EDITED BY

J. MANSFELD, D.T. RUNIA
AND J.C.M. VAN WINDEN

31. EDLOW, R.B. *Galen on Language and Ambiguity.* An English Translation of Galen's *De Captionibus* (On Fallacies), With Introduction, Text and Commentary. 1977. ISBN 90 04 04869 3

34. EPIKTET. *Vom Kynismus.* Herausgegeben und übersetzt mit einem Kommentar von M. Billerbeck. 1978. ISBN 90 04 05770 6

35. BALTES, M. *Die Weltentstehung des platonischen Timaios nach den antiken Interpreten.* Teil 2. Proklos. 1979. ISBN 90 04 05799 4

37. O'BRIEN, D. *Theories of Weight in the Ancient World.* Four Essays on Democritus, Plato and Aristotle. A Study in the Development of Ideas 1. Democritus: Weight and Size. An Exercise in the Reconstruction of Early Greek Philosophy. 1981. ISBN 90 04 06134 7

39. TARÁN, L. *Speusippus of Athens.* A Critical Study with a Collection of the Related Texts and Commentary. 1982. ISBN 90 04 06505 9

40. RIST, J.M. *Human Value.* A Study in Ancient Philosophical Ethics. 1982. ISBN 90 04 06757 4

41. O'BRIEN, D. *Theories of Weight in the Ancient World.* Four Essays on Democritus, Plato and Aristotle. A Study in the Development of Ideas 2. Plato: Weight and Sensation. The Two Theories of the 'Timaeus'. 1984. ISBN 90 04 06934 8

44. RUNIA, D.T. *Philo of Alexandria and the Timaeus of Plato.* 1986. ISBN 90 04 07477 5

45. AUJOULAT, N. *Le Néo-Platonisme Alexandrin: Hiéroclès d'Alexandrie.* Filiations intellectuelles et spirituelles d'un néo-platonicien du Ve siècle. 1986. ISBN 90 04 07510 0

46. KAL, V. *On Intuition and Discursive Reason in Aristotle.* 1988. ISBN 90 04 08308 1

48. EVANGELIOU, CH. *Aristotle's Categories and Porphyry.* 1988. ISBN 90 04 08538 6

49. BUSSANICH, J. *The One and Its Relation to Intellect in Plotinus.* A Commentary on Selected Texts. 1988. ISBN 90 04 08996 9

50. SIMPLICIUS. *Commentaire sur les Catégories.* Traduction commentée sous la direction de I. Hadot. I: Introduction, première partie (p. 1-9, 3 Kalbfleisch). Traduction de Ph. Hoffmann (avec la collaboration d'I. et P. Hadot). Commentaire et notes à la traduction par I. Hadot avec des appendices de P. Hadot et J.-P. Mahé. 1990. ISBN 90 04 09015 0

51. SIMPLICIUS. *Commentaire sur les Catégories.* Traduction commentée sous la direction de I. Hadot. III: Préambule aux Catégories. Commentaire au premier chapitre des Catégories (p. 21-40, 13 Kalbfleisch). Traduction de Ph. Hoffmann (avec la collaboration d'I. Hadot, P. Hadot et C. Luna). Commentaire et notes à la traduction par C. Luna. 1990. ISBN 90 04 09016 9

52. MAGEE, J. *Boethius on Signification and Mind.* 1989. ISBN 90 04 09096 7

53. BOS, E.P. and MEIJER, P.A. (eds.) *On Proclus and His Influence in Medieval Philosophy.* 1992. ISBN 90 04 09429 6

54. FORTENBAUGH, W.W., et al. (eds.) *Theophrastes of Eresos.* Sources for His Life, Writings, Thought and Influence. 1992. ISBN 90 04 09440 7 *set*

55. SHANKMAN, A. *Aristotle's* De insomniis. A Commentary. ISBN 90 04 09476 8

56. MANSFELD, J. *Heresiography in Context.* Hippolytos' *Elenchos* as a Source for Greek Philosophy. 1992. ISBN 90 04 09616 7

57. O'BRIEN, D. *Théodicée plotinienne, théodicée gnostique.* 1993. ISBN 90 04 09618 3

58. BAXTER, T.M.S. *The Cratylus.* Plato's Critique of Naming. 1992. ISBN 90 04 09597 7

59. DORANDI, T. (Hrsg.) *Theodor Gomperz. Eine Auswahl herkulanischer kleiner Schriften (1864-1909).* 1993. ISBN 90 04 09819 4

60. FILODEMO. *Storia dei filosofi. La stoà da Zenone a Panezio* (PHerc. 1018). Edizione, traduzione e commento a cura di T. Dorandi. 1994. ISBN 90 04 09963 8

61. MANSFELD, J. *Prolegomena*. Questions to be Settled Before the Study of an ✗
    Author, or a Text. 1994. ISBN 90 04 10084 9
62. FLANNERY, S.J., K.L. *Ways into the Logic of Alexander of Aphrodisias*. 1995.
    ISBN 90 04 09998 0
63. LAKMANN, M.-L. *Der Platoniker Tauros in der Darstellung des Aulus Gellius*. 1995.
    ISBN 90 04 10096 2
64. SHARPLES, R.W. *Theophrastus of Eresus*. Sources for his Life, Writings, Thought
    and Influence. Commentary Volume 5. Sources on Biology (Human Physiol-
    ogy, Living Creatures, Botany: Texts 328-435). 1995. ISBN 90 04 10174 8
65. ALGRA, K. *Concepts of Space in Greek Thought*. 1995. ISBN 90 04 10172 1   ✗
66. SIMPLICIUS. *Commentaire sur le manuel d'Épictète*. Introduction et édition critique
    de texte grec par Ilsetraut Hadot. 1995. ISBN 90 04 09772 4
67. CLEARY, J.J. *Aristotle and Mathematics*. Aporetic Method in Cosmology and ✗
    Metaphysics. 1995. ISBN 90 04 10159 4
68. TIELEMAN, T. *Galen and Chrysippus on the Soul*. Argument and Refutation in the
    De Placitis Books II-III. 1996. ISBN 90 04 10520 4
69. HAAS, F.A.J. DE. *John Philoponus' New Definition of Prime Matter*. Aspects of its
    Background in Neoplatonism and the Ancient Commentary Tradition. 1997.
    ISBN 90 04 10446 1
71. ANDIA, Y. DE. *Henosis*. L'Union à Dieu chez Denys l'Aréopagite. 1996. ISBN
    90 04 10656 1
72. ALGRA, K.A., HORST, P.W. VAN DER, and RUNIA, D.T. (eds.) *Polyhistor*. ✗
    Studies in the History and Historiography of Ancient Philosophy. Presented to
    Jaap Mansfeld on his Sixtieth Birthday. 1996. ISBN 90 04 10417 8
73. MANSFELD, J. and RUNIA, D.T. *Aëtiana*. The Method and Intellectual Con-
    text of a Doxographer. Volume 1: The Sources. 1997. ISBN 90 04 10580 8
74. SLOMKOWSKI, P. *Aristotle's* Topics. 1997. ISBN 90 04 10757 6
75. BARNES, J. *Logic and the Imperial Stoa*. 1997. ISBN 90 04 10828 9
76. INWOOD, B. and MANSFELD, J. (eds.) *Assent and Argument*. Studies in Cicero's
    Academic Books. Proceedings of the 7th Symposium Hellenisticum (Utrecht,
    August 21-25, 1995). 1997. ISBN 90 04 10914 5
77. MAGEE, J. (ed., tr. & comm.) *Anicii Manlii Severini Boethii* De divisione liber.
    Critical Edition, Translation, Prolegomena, and Commentary. 1998.
    ISBN 90 04 10873 4
78. OLYMPIODORUS. *Commentary on Plato's* Gorgias. Translated with Full Notes
    by R. Jackson, K. Lycos, & H. Tarrant. Introduction by H. Tarrant. 1998.
    ISBN 90 04 10972 2
79. SHARPLES, R.W. *Theophrastus of Eresus*. Sources for his Life, Writings, Thought
    and Influence. Commentary Volume 3.1. Sources on Physics (Texts 137-223).
    With Contributions on the Arabic Material by Dimitri Gutas. 1998.
    ISBN 90 04 11130 1
80. MANSFELD, J. *Prolegomena Mathematica*. From Apollonius of Perga to Late
    Neoplatonism. With an Appendix on Pappus and the History of Platonism.
    1998. ISBN 90 04 11267 7

4 n.10   Arist. EN 1095a 11

8   beyond my competence

9   ὁ καλούμενος ἀναλυόμενος

29   Schol V. No1 (on Book V of Elements) 'some say ... Eudoxus'
      See also pp.126-7 & Complementary Note 108 on Scholion V. No3

27   Sch. VII. No3 on monad.   28 .n.104   Query - Scholia in Euclidem)
                                                        in Barbera?

                                                              Cp. p 33

32   Pappus' Platonism

43 n.143   Eutocius' date ... the school at Alexandria Cp 48

44 n.145   career of Apollonius

46 n.151   Knorr's 'book within a book'

48   Vita of Archimedes

49   'he observed' (Heron)

66 n.226   Philosophical interests of Ptolemy

68 'a clearly Platonic touch' n 230 Nicomach on numeric ratios

82   Nicomachus. See also p.19

86-7 'ancients' ~ νεώτεροι

88   Ammonius Hermiae (Above p. 43)

93   Pappus on philosophers & mathematicians. Heron. Late
 -94 | 99-100 'Middle Platonism' 'Neoplatonism'   antiquity

115   Proclus refers reader to Euclid & Archimedes, n. 382

114 n. 394 πυθαγορίζει

117-121 means, or proportions. See also Index sub proportion(s)